Praise for *Ag*

"*Against Doom* lays out key elements of a far-reaching, global-scaled, pragmatic, people-powered strategy to topple the power of the fossil fuel industry and the institutions behind it."

—David Solnit, author of *Globalize Liberation: How to Uproot the System and Build a Better World*

"In *Against Doom*, Brecher has provided the climate movement with two essential tools: a moral framework for the struggle against fossil fuels, and an actual plan for victory. By blending sober social movement analysis with the fire of grassroots activism, this book shows that there is a genuine, and winnable, case against the fossil fuel economy—a case to be argued in the streets as well as the courtroom. It's an essential volume for anyone committed to social change in the fight against climate change."

—Joseph Hamilton, Climate Defense Project

"Jeremy Brecher has managed to combine his passion for justice with the wisdom gleaned from decades of research, writing, and personal engagement in social movements and condense them into this short and accessible strategy document. Having benefited from his guidance on a variety of campaigns over the last twenty years, I look forward to collaborating with other organizers and activists in using this book as a blueprint for building a global climate insurgency."

—John Humphries, Connecticut Roundtable on Climate and Jobs

"A crisp, clear, and savvy synthesis of key concepts and ideas that will help the global climate justice movement to succeed. Brecher outlines many feasible climate solutions that should give all of us hope, despite the odds."

—Naomi Klein, author of *This Changes Everything: Capitalism vs. the Climate*

AGAINST
DOOM

TO DEFEND THE CLIMATE, THE CONSTITUTION, AND THE PUBLIC TRUST

We are here to help our community, our country, and the world BREAK FREE FROM FOSSIL FUELS.

While we may risk arrest, we commit no crime.

Today the fossil fuel industry and the governments that do its bidding are laying waste to the earth's atmosphere, the common property of humanity.

We are upholding fundamental principles embodied in the laws and constitutions of countries around the world.

We are upholding the unalienable rights to life and liberty.

We are implementing the public trust doctrine, which requires that vital natural resources on which human well-being depend must be cared for by our governments for the benefit of all present and future generations.

Governments have no right to authorize the destruction of the public trust.

Governments have no right to wreck the rights to life and liberty for future generations.

We are here to enforce the law on governments and corporations that are committing the greatest crime in human history.

Those who take nonviolent direct action—blockade coal-fired power plants or sit down at the White House to protest fossil fuel pipelines—are exercising their fundamental constitutional rights to life and liberty and their responsibility to protect the atmospheric commons they own along with all of present and future humankind.

WE PROCLAIM: The people of the world have a right, indeed a duty, to protect the public trust we own in common—the earth's climate.

When we take nonviolent direct action we are law-enforcers carrying out our duty to protect the earth's climate from illegal, dangerous crimes.

PUBLIC TRUST
CLIMATE PROTECTION
ENFORCEMENT

AGAINST
DOOM

A Climate Insurgency Manual

Jeremy Brecher

Against Doom: A Climate Insurgency Manual
Jeremy Brecher

PM Press
PO Box 23912
Oakland, CA 94623
www.pmpress.org

Cover design by John Yates/stealworks.com
Layout by Jonathan Rowland

ISBN: 978-1-62963-385-5
Library of Congress Control Number: 2016959585

10 9 8 7 6 5 4 3 2 1

Printed in the USA by the Employee Owners of Thomson-Shore
in Dexter, Michigan.
www.thomsonshore.com

Contents

I would like to thank my current and past colleagues at the Labor Network for Sustainability, Joe Uehlein, Becky Glass, Michael Leon Guerrero, Harley Mocker, Brendan Smith, and the late Tim Costello, who have nurtured my work on climate protection.

Thanks also to the members of the Break Free from Fossil Fuels Public Trust Working Group.

John Humphries and Todd Vachon have reviewed and made suggestions for parts of this book.

Jill Cutler has once again provided me her services as "in-house editor."

Prologue
Climate Insurgency vs. Trump

BEFORE THE ELECTION OF DONALD TRUMP AS PRESIDENT of the United States, the world was speeding toward climate catastrophe—while world leaders opined that it would be a good idea to slow things down. Now President Trump has jammed his foot on the global warming accelerator. The consequences will be as predictable as they will be horrendous. Is there any way for the rest of us to put on the brakes?

Trump has threatened to ignore or even withdraw from the Paris climate agreement. He has alleged that global warming is a fraud perpetrated by the Chinese to steal American jobs. He has pledged to expand fossil fuel extraction to the max. His plans will heat up the earth far beyond the 6.3 degrees Fahrenheit increase scientists expect even if all countries meet their current climate pledges.

Since scientists first established that greenhouse gases (GHGs) from fossil fuels were causing the earth to heat up, climate protection advocates have sought to persuade governments to sharply reduce the burning of fossil fuels. But Trump is committed to greatly expanding fossil fuel production and use—radically increasing GHG emissions.

Conventional means of lobbying and persuading the powerful are likely to have little impact on his agenda. Climate insurgency provides another way of putting on the brakes.

Climate insurgency is a strategy for using the power of the people to realize our common interest in protecting the climate. It uses mass, global, nonviolent action to challenge the legitimacy of the public officials in the U.S. and elsewhere who are perpetrating climate destruction. It has the potential to halt and roll back Trump's fossil fuel agenda and the global thrust toward climate destruction.

In the Trump era, new laws, executive orders, and agency rulings will no doubt authorize a huge development of mines, wells, pipelines, power plants, and other new fossil fuel infrastructure. Their supporters will claim that all this is being done according to law—and will no doubt produce permits issued by the Trump administration to prove it. They will even accuse those who are nonviolently opposing them of being criminals and thugs (as peaceful Native American water protectors were criminalized when they opposed the Dakota Access Pipeline at Standing Rock). Those accusations will be used to justify harassment, legal and extralegal violence, arrest, prosecution, and incarceration.

Climate insurgency maintains that however many permits with official signatures and seals the Trump administration may issue, nothing gives them the legal authority to destroy the earth's climate and thereby all those who depend on it. As a recent federal court decision put it, "the right to a climate system capable of sustaining human life is fundamental to a free and ordered society."[1]

Trump's expansion of fossil fuel infrastructure and thereby of GHG emissions is illegitimately imposing destruction on the earth's climate. It poses an existential threat to the world's people. We the people have a right, indeed an obligation, to resist.

Climate destruction is only one consequence of the horrifying Trump agenda. Trump's presidency threatens immigrants, African Americans, Muslims, workers, women, children, the elderly, the disabled, LGBTQ people, and many others. The day Trump was elected, demonstrations broke out against him in dozens of American cities; police chiefs, mayors, and governors declared they would not implement his attack on immigrants; thousands of people signed up to accompany threatened immigrants, religious minorities, and women.

Climate protection is part of this wider resistance to Trump's antisocial agenda of destroying people and the planet for the aggrandizement of those who hope to gain from that destruction. Climate insurgency can be a critical element in unifying resistance to that agenda. First, because climate change hurts all of us, creating a common interest in climate protection. Second, because a transition to a climate-safe economy can provide the transformation we need to establish a more just country and world.

It is impossible to know what the future course of the Trump era will be. We do know that the future of the planet and its people depend on resisting and overcoming Trump's agenda. If there are future historians, Trump and Trumpism will be remembered as a failed attempt to prevent humanity from protecting itself against climate catastrophe through a global nonviolent insurgency.

Introduction
Climate Insurgency vs. Doom

UNTIL THE ELECTION OF DONALD TRUMP AS PRESIDENT OF the United States, world leaders projected confidence that their countries were on the road to fixing the earth's climate. With the world's historically greatest greenhouse gas (GHG) polluter now ruled by climate change deniers, coal and oil executives, and fossil fuel promoters, we cannot count on governments to protect our climate. Now climate protection is up to the people. The only way to combat devastating climate change may be climate insurgency.

The illusion that world leaders were fixing climate change was exemplified by the 2015 Paris Agreement in which 195 countries acknowledged their individual and collective duty to protect the earth's climate—and willfully refused to perform that duty. They unanimously agreed to the goal of keeping global warming "well below 2 degrees Celsius" and of pursuing efforts "to limit the increase in temperatures to 1.5 degrees Celsius." But they did not agree to a single legally binding requirement about how, or how much, they would cut emissions.[1] Consequently it is now up to the people of the world to halt climate destruction. *Against Doom: A Climate Insurgency Manual* tells how.

In response to climate destruction we are witnessing the birth of a global nonviolent constitutional insurgency. Global because the world order of climate destruction it seeks to change is global. Nonviolent because it relies on the power of the world's people to withdraw our acquiescence and cooperation from those who are destroying our planet. Constitutional because it is based on fundamental constitutional principles: people have a right to a climate system capable of sustaining human life, the earth's shared resources belong to the people, and therefore governments have no authority to destroy them.

In a global survey by the Pew Research Center, a majority of those surveyed say climate change is a very serious problem and nearly four in five say their countries should limit greenhouse gas emissions as part of an international agreement.[2] A majority believe people are already being harmed by climate change; an additional 28% think people will be harmed in the next few years; a majority are concerned that it will cause harm to them personally during their lifetime; and 40% are very worried this will happen.[3]

Many of them are ready to act. One in six Americans say they would personally engage in nonviolent civil disobedience against corporate or government activities that make global warming worse. That's about forty million adults.[4] The fate of the earth may depend on them—and others around the world—doing so.

Although the people of the world want to protect the climate, the governments and fossil fuel corporations of the world go on destroying it. The climate insurgency is a vehicle for the world's people to realize our common interest and our goal of restoring balance to the earth's climate system. Through mass demonstrations, meetings, and marches, through civil disobedience, and perhaps even through popular tribunals and other forms of countergovernment, the climate insurgency can use the power of the world's people to begin forcing climate protection on the powers and principalities of the earth. Will it be able to do so? As Gandhi once said of India's struggle to free itself from British rule, "The matter resolves itself into one of matching forces."

Part I, "The Global Climate Insurgency," explains how and why the climate insurgency has emerged. Chapter 1, "This Is What Insurgency Looks Like," tells the story of Break Free from Fossil Fuels, the May 2016 global weeks of action that constituted the largest global civil disobedience against fossil fuels up to that time. Chapter 2, "Paris: The Promise of Betrayal" dissects the Paris climate agreement, the charade which purports to counter global warming but actually authorizes unlimited climate destruction. Chapter 3, "The Power of the People vs. the Forces of Doom" explains the strategy underlying the emerging global climate insurgency.

Part II, "Climate Insurgency in America," lays out a strategy for climate insurgents in the United States. Chapter 4, "A Strategic Vision," assesses the strengths and weaknesses of climate destroyers and climate protectors and analyzes the underlying dynamics of their struggle. Chapter 5, "Imposing a Fossil Freeze" and Chapter 6, "Imposing Climate Action Plans," describe two essential actionable objectives through which the climate insurgency can force a change in energy production and use. Chapter 7, "Self-Organization for Climate Defense," describes the climate insurgency's emerging organizational form.

Chapter 8, "Turning the Public against Fossil Fuels," and Chapter 9, "Turning Climate Worriers Into Climate Warriors," explain how public opposition, confusion, and apathy can be turned into action. Chapter 10, "Undermining the Pillars of Support for Climate Destruction," tells how the power of the fossil fuel forces can be weakened to the point where it crumbles. Chapter 11, "Just Transitions," outlines a program through which the climate insurgency can ensure wide benefits from the transition to a climate-safe economy while protecting those the transition might threaten.

Chapter 12, "The Right of the People to Protect the Climate," lays out how the public trust doctrine and other constitutional principles empower the climate insurgents to proclaim that they—not those who are destroying the climate—embody the law. Chapter 13, "Dual Power," explains the ultimate weapon of the climate insurgency: the nonviolent withdrawal of cooperation and consent to the illegitimate authorities who protect and perpetuate climate destruction. The Conclusion, "Two Scenarios," summarizes what will happen if the climate insurgency loses and how it might succeed.

My 2015 book *Climate Insurgency: A Strategy for Survival* (available for free download at www.jeremybrecher. org) recounted the failure of climate protection over the past quarter century and laid out a broad strategy for overcoming it through a global nonviolent constitutional insurgency. *Against Doom: A Climate Insurgency Manual* tells how to put that strategy into action—and how it can succeed. It is a handbook for halting global warming and restoring our climate—a how-to for climate insurgents.

I write as an advocate for and a participant in the climate insurgency. I have been writing about the climate protection movement since 1988; I helped found the Labor Network for Sustainability; I was arrested at the White House in the early protests against the Keystone XL pipeline; I helped block the train tracks in Albany during Break Free from Fossil Fuels. I repeat here what I wrote at the end of the Introduction to *Climate Insurgency*:

> I hope readers will examine the strategy proposed here critically. But I also hope they will either correct its flaws or develop a better alternative. Climate protection can't wait for a perfect strategy; all of us have a duty to find the best strategy—and then act on it.

PART I
THE GLOBAL
CLIMATE
INSURGENCY

Chapter 1
This Is What Insurgency Looks Like

In a small church in Albany, New York's low-income, predominantly AfricanAmerican South End, forty people were gathered for a community meeting. They were organizing a protest against trains carrying potentially explosive oil—dubbed by the residents "bomb trains"—that were running through their neighborhood. City Counselor Vivian Kornegay told the group that many municipalities had opposed the bomb trains and other dangerous fossil fuel infrastructure but had little power to protect their residents; it was up to a "people's movement" to do so. "What we want is for all of us to be free, healthy, and safe—and for our planet to be a better place to live."

Maeve McBride, an organizer for 350.org, explained that the protest was part of a global campaign of direct action and civil disobedience aiming to keep 80% of all fossil fuels in the ground. Pastor Mark Johnson of the St. John's Church of God in Christ said, "I heard at a meeting last night that we have a constitutional right to clean water and clean air." Maeve McBride explained that the action was part of a "new wave" that was drawing on a "new paradigm"—"using civil disobedience to protect the public trust," which includes water, air, and the climate itself.

Organizers had met with officials from the police and sheriff's offices and reported, "They abhor the trains—and are very supportive of us." Then the group received direct action training. They read out loud the "action agreement" pledging nonviolent behavior and mutual support. Then they lined up to march and while police officers (played by the trainers) ordered them to move away, they scrambled onto an imaginary railroad track. Later that evening the steering committee for Albany Break Free planned outreach to supporting organizations, phone banks, canvassing, leafleting, and details of the action.

The Albany organizers had learned about the "new paradigm" when 350.org, North American co-organizers of Break Free from Fossil Fuels, had decided to use the "public trust" principle to frame U.S. Break Free actions and formed a Break Free Public Trust Work Group to spread the idea. Some on the Break Free Albany steering committee had participated in the working group's webinar on using the public trust doctrine, and they decided to integrate the Public Trust Proclamation into their "topline message" and to hand out the Break Free Public Trust Proclamation to all participants. [See Frontispiece for Break Free Public Trust Proclamation.]

A week before the action the Break Free Albany steering committee defined their basic message. Potentially explosive crude oil bomb trains roll through Albany and surrounding communities, polluting the air and contributing to the climate crisis. Primarily low-income communities of color are put at risk. The urgent need to address climate change means that fossil fuels have to be left in the ground and a transition made to a "twenty-first century renewable energy economy." They called for an end to all new fossil fuel infrastructure, including pipelines, power plants, compressor stations, and storage tanks. And they called for a just transition away from fossil fuel energy with training and jobs for affected workers, so "no worker is left behind."

On Six Continents

Meanwhile, reports of Break Free actions on six continents began flowing in. In Wales, protesters shut down the UK's largest open-pit coal mine for over twelve hours with no arrests. In the Philippines, ten thousand people marched and rallied demanding the cancellation of a six-hundred-megawatt coal power plant project. In New Zealand, protesters blockaded and shut down Christchurch, Dunedin, and Wellington branches of the ANZ bank, which had $13.5 billion invested in fossil fuels. In Indonesia, banner drops brought a coal terminal to a standstill, and three thousand people held a "climate carnival" at the presidential palace demanding a move from coal to renewable energy. In Germany, four thousand people shut down a large lignite coal mine for more than two days.

In Australia, two thousand people shut down the world's largest coal port with a kayak flotilla and a railroad blockade. In Brazil, thousands participated in a protest against fracking during a concert at an annual rural fair. In Nigeria, demonstrations called attention to the environmental and social devastation that followed in the wake of exhausted oil wells. In South Africa, drought-affected farmers and communities from around the country came together for a "speak out and bread march." In Ecuador, activists planted trees on the future site of an oil refinery to protest drilling in a national park. In Vancouver, Canada, more than eight hundred people held a sit-in and a kayak swarm at the tanker terminal for the Kinder Morgan gas pipeline. In Turkey, community leaders led a mass action at a coal waste site calling for a halt to four fossil fuel plant projects planned for the area.

Outside Seattle, thousands converged on two oil refineries with kayak flotillas, a march was led by Indigenous leaders, and an overnight sit-in on the train tracks led to more than fifty arrests. In Washington, DC, 1,300 demanded no new offshore drilling in the Arctic and off the Gulf Coast. Outside Chicago, dozens were arrested as a thousand people protested a planned expansion of a BP refinery. In Los Angeles, two thousand opposed the oil drilling that is conducted right within the city. In Lakewood, Colorado, hundreds of people delayed an auction of thousands of acres of public land for oil and gas drilling with disruptions and a sit-in. Organizers called Break Free "the largest ever global civil disobedience against fossil fuels."[1]

Creative Tension

In some cases, Break Free evoked what Martin Luther King Jr. characterized in his "Letter from a Birmingham Jail" as "creative tension." Some early Break Free statements in the state of Washington, for example, suggested that protesters might use direct action to shut down oil refineries. This was understandably alarming to workers in the highly dangerous refineries. Break Free organizers and a retired union official initiated discussions with the local union that made clear that Break Free would not try to obstruct the plants or their workers and that took into consideration other safety concerns of the local union.

Break Free had always advocated a "just transition," but discussions with the local union helped them better understand what that means from the workers' point of view. Break Free organizers say they came away committed to educating their constituency about the importance of fighting to protect and create family-wage jobs in the transition to a clean energy economy; protecting job security amidst declines in fossil fuel consumption; and minimizing job losses as the necessary action is taken to curtail dangerous climate change.

Notwithstanding this dialogue between local Break Free organizers and the union local, the national United Steelworkers union issued a statement critical of 350.org and Break Free. Noting that three USW-represented oil refineries were targeted locations, the USW said, "shutting down a handful of refineries in the U.S. would likely lead to massive job loss in refinery communities, increased imports of refined oil products, and ultimately no impact on reducing global carbon emissions." It added that "short-sighted and narrow-focused activities like 350.org's 'Break Free' actions this May make it much more challenging to work together to envision and create a clean energy economy." But they added, "The work of addressing climate change and building a more sustainable economy is too important to be derailed by a handful of groups organizing protests at our plant gates."[2]

As Dr. King wrote, "Nonviolent direct action seeks to create such a crisis and establish such creative tension that a community that has constantly refused to negotiate is forced to confront the issue." While Break Free and the Steelworkers by no means see eye to eye, they have begun to negotiate, and are even discussing cooperation around upcoming local energy issues.

Personal and Global

In Albany, a "climate camp" made preparations for the action, creating banners and other artwork in an "artbuild," organizing logistics, and nailing down final plans. As Chairman Norman Bay of the Federal Energy Regulatory Commission addressed the Independent Power Producers of New York, a group of Break Free protesters interrupted and drove him from the stage. "Why are you signing the death warrant for so many people?" one protester asked Bay. IPPNY president Gavin Donahue said the protesters were "aggressive," "disruptive," and "out of line." "We were keeping an eye out, but we did not see anything before this," he said. "The protesters disguised themselves in suits and ties to blend in."[3]

As the sun set over the Hudson River the evening before the action, a kayak flotilla provided a perfect photo-op. That night hundreds from the Northeast gathered in a historic neighborhood black church for a meal, a rally, and civil disobedience training. They shared stories from struggles to block pipelines, fracking, and other fossil fuel projects around the region and celebrated the state's refusal of permits for the Constitution Pipeline after massive protests.

Saturday morning more than 1,500 people arrived at Lincoln Park. Rev. Mark Johnson, welcoming people to the park where he had played as a child, said, "We all deserve clean water, we all deserve clean air." Albany Common Councilmember Vivian Kornegay said, "We're tired of big oil coming to our communities and polluting. We should keep the oil in the ground and make our country a greener and safer place. My community is in danger. The people in the Ezra Prentice housing are facing asthma and cancer. The company says it wants to partner with us, but it is partnership where we assume all the risks and minimum benefits." Miss Charlene Benton, president of the Ezra Prentice Homes Tenants' Association said, "We're not going to be cremated without permission."[4]

Attorney Mark Mishler observed, "Right near here is a historic mansion possessed by slave owners. Slavery was legal. That didn't make it right. The bomb trains are not legal because they are not right." Rev. Johnson added, "We're going to stay on the right side of the law because the moral side of the law is the right side of the law."

Then Rev. Johnson read the Break Free Albany "Action Agreements." Participants agreed "not to harm people or property"; to be "dignified in dress, demeanor, and language";

to attend action training; and to act "in accordance with our group's agreed plan for action."

As the crowd started to march toward the Port of Albany, 500 of the demonstrators peeled off to show their support for the people of the Ezra Prentice Homes, a 179-unit housing development described by the protesters as "ground zero for environmental racism," where bomb trains run next to the playground and the railroad parks its trains free of charge.

As the marchers reached the train tracks, police were present but stood by as they occupied the tracks. An organizer hollered, "You came to block the tracks and that is exactly what you are doing."

As people settled in on the tracks, law student Kelsey Skaggs asked those risking arrest to fill out intake forms for the legal team. Then she said:

> But I want to talk about a different kind of lawbreaking. Fossil fuel companies, and the governments that authorize their activities, are destroying the land, water, and atmosphere that sustain us. From my home in Alaska to here in Albany, we are being sacrificed for the profit of fossil fuel companies.
>
> But these companies have a problem. Their problem is that we—all of us—have rights to that land, water, and atmosphere.
>
> We have rights under a legal principle called the public trust. The public trust concept is old law—it's been around since ancient Rome. In American law, it means that the government has a duty to protect shared natural resources, and to hold them in trust for the public and for future generations.
>
> But our governments are violating this obligation by failing to regulate fossil fuel emissions. They violate

this right by subsidizing fossil fuels, by approving new dirty energy projects, and by locking us into further, deadly emissions.

The question—the critical question facing humanity at this moment—is what we are going to do about these violations of our rights.

In the face of government's failures, there is only one answer that leads to a livable future on this planet. And that answer is: what each of you is doing, right here, today. Standing up and taking action to break free and end the era of fossil fuels. Enforcing the public trust. Demanding that the government fulfill its obligations to the people, not bow to corporate power.

It's up to us to claim our right to a healthy climate, to stand up for the rights of future generations, and to stop the degradation of our planet and our communities and everything that we love. Thank you for doing that.

More than 400 of the 1,500 people registered for the action said they would be willing to be arrested for physically blocking the trains.[5] The company had canceled trains through Albany for the day because of the protest. But sixteen miles up the track in Guilderland, climate activists Marissa Shea and Maeve McBride suspended themselves from train tracks on a railroad bridge. At the risk of a serious or even fatal accident they blockaded a bomb train carrying fracked crude oil from North Dakota. They and three members of their support team were arrested after successfully delaying the train.

Shea and McBride described their efforts as enforcing the public trust doctrine which requires that vital natural resources, in this case the atmosphere, on which human well-being depends, must be cared for by our governments for the benefit of present and future generations.

"The global climate system, on which every human depends, is no longer stable because our governments have utterly failed us. So now, for our survival, we will act on climate ourselves," said Shea.

The activists demand that the business as usual economy, which is currently reliant on fossil fuels, must be transformed into a new fossil free economy that is just and equitable, a just transition.

"Most of my family lives within a few miles of where the bomb trains travel. This is personal and global. Their lives are at risk and millions of lives are at risk with rising seas, forest fires, violent storms, and all the havoc that global warming brings," said McBride, who grew up in Troy. "Today I felt called to directly obstruct the fossil fuel industry, joining thousands of others around the world."

McBride had earlier written Break Free organizers around the country:

> Many of us participated in the Public Trust webinar a couple of weeks ago, and we are excited for the paradigm-shifting opportunity that this presents. Organizers have asked those risking arrest to consider a court solidarity approach where, as a group, we will plead not-guilty and seek to bring our cases to court. While the DA is likely to drop the majority of charges, we are taking measures to ensure that we will have some viable court cases and will seek to argue them under the Public Trust Doctrine and/or necessity with a Public Trust spin. During the action we will distribute copies of the Break Free Public Trust Proclamation, as it could be important to have this document in hand during arrests [and for some political theater in court].[6]

The call to Break Free from Fossil Fuels envisioned "tens of thousands of people around the world rising up" to take back control of their own destiny; "sitting down" to "block the business of government and industry that threaten our future"; conducting "peaceful defense of our right to clean energy." That's just what happened.

Such a "rising up" amounts to a global nonviolent insurgency—a withdrawal of consent from those who claim the right to rule—manifested in a selective refusal to accept and obey their authority.[7] Break Free from Fossil Fuels represented a quantum leap in the emergence of a global nonviolent climate insurgency—its nonviolent "shot heard around the world." It was globally coordinated, with common principles, strategy, planning, and messaging. It used nonviolent direct action not only as an individual moral witness but also to express and mobilize the power of the people on which all government ultimately depends. It presented climate protection not only as a moral but as a legal right and duty, necessary to protect the Constitution and the earth's essential resources on which we and our posterity depend. It represented an insurgency because it denied the right of the existing powers and principalities—be they corporate or governmental—to use the authority of law to justify their destruction of the earth's climate.

Chapter 2
Paris: The Promise of Betrayal

BREAK FREE FROM FOSSIL FUELS WAS IN CONSIDERABLE part a response to the Paris Agreement on climate change. But what is the Paris Agreement?

The 195 nations meeting in Paris in December 2015 unanimously agreed to the goal of keeping global warming "well below 2 degrees Celsius" and to pursue efforts "to limit the increase in temperatures to 1.5 degrees Celsius." Despite that goal, The Paris climate agreement doesn't prevent one molecule of greenhouse gas (GHG) from being put in the atmosphere; indeed, it permits every country in the world to increase the GHG emissions that cause global warming without limit. The U.S. Department of Energy now predicts that global GHG emissions will increase 40% by 2040.[1]

Under the Paris Agreement, governments put forward any targets they want—known as Intended Nationally Determined Contributions (INDCs)—with "no legal requirement dictating how, or how much, countries should cut emissions."[2] These voluntary commitments don't come into effect until 2020 and generally end in 2025–2030.

Today there are 400 parts per million (ppm) of carbon in the atmosphere, far above the 350 ppm climate scientists regard as the safe upper limit. Even in the unlikely event that all nations fulfill their INDC pledges, carbon in the atmosphere is predicted to increase to 670 ppm by the end of this century.[3] The global temperature will rise an estimated 3.5°C (6.3°F) above preindustrial levels.[4] For comparison, a 1°C increase was enough to cause all the effects of climate change we have seen so far, from Arctic melting to desertification. In short, the agreement authorized the continued and even increased destruction of the earth's climate.[5]

U.S. negotiators were adamant that the agreement must not include any binding restrictions on emissions. Secretary of State John Kerry told fellow negotiators he "wished that we could include specific dates and figures for emissions cuts and financial aid to developing countries," but "this could trigger a review by the US Senate that could scuttle the entire agreement."[6] When U.S. lawyers discovered a phrase declaring that wealthier countries "shall" set economy-wide targets for cutting their GHG pollution, Kerry said, "We cannot do this and we will not do this. And either it changes or President Obama and the United States will not be able to support this agreement." "Shall" was changed to "should" without so much as a vote.[7]

The breathtaking gap between the Paris Agreement's aspiration to hold global warming below 2°C and the agreement's actual commitments is indicated by an analysis by Climate Interactive and MIT Sloan. The current U.S. pledge to drop GHG emissions 26% below 2005 levels by 2025, along with the pledges of other countries, will lead to a global temperature increase of 3.5°C (6.3°F) above preindustrial levels. To reduce warming to 1.8°C (3.2°F) will require the U.S. to increase its INDC from 26% below 2005 levels to 45% by 2025, and for other countries to make comparable reductions.[8]

Under the Paris Agreement countries will monitor their emissions and reconvene every five years starting in 2023 to report on the results and perhaps ratchet up their INDCs. This has been characterized as creating a "name-and-shame" system of global peer pressure, "in hopes that countries will not want to be seen as international laggards."[9]

On the last day of the Paris Summit, a panel of leading climate scientists evaluated what would be necessary to achieve its targets. Prof. Hans Joachim Schellnhuber of the Potsdam Institute for Climate Impact Research said that to reach the 2°C target the world would have to get CO_2 out of its system by 2070. To reach the 1.5°C target it would have to eliminate CO_2 emissions by 2050. Johan Rockstrom of the Stockholm Resilience Center said that for any chance of reaching 1.5°C, the richest nations need to reach zero fossil fuel use by 2030.[10]

In Paris, the governments of the world made a promise to the people of the world—and immediately betrayed it. On the one hand, the nations of the world, including the U.S., agreed that "climate change represents an urgent and potentially irreversible threat to human societies and the planet" and that "deep reductions in global emissions will be required." On the other hand, they simply accepted "the significant gap" between the "aggregate effect of Parties' mitigation pledges in terms of global annual emissions of greenhouse gases by 2020" and "aggregate emission pathways consistent with holding the increase in the global average temperature to well below 2°C above preindustrial levels and pursuing efforts to limit the temperature increase to 1.5°C."[11] Stripped of the jargon, this says they recognize that their current actions are aggravating that "irreversible threat" and acknowledge the inadequacy of their efforts to halt it.

The governments of the world may rule the world but they don't own the world—that is the common property of humanity. The Paris Summit was in effect a conspiracy of the world's governments to rob the world's people and our posterity of our common inheritance. Nations acknowledged the devastation they are causing but refused to be accountable to each other for correcting it. So now they need to be made accountable to the world's real owners.

Chapter 3

The Power of the People vs. the Forces of Doom

Climate protection is unquestionably in the interest of virtually all of the world's people. Yet we have been unable to impose that interest on the world. In a fairytale version of how the world works, the people of each country should be able to elect governments that send delegates to international climate negotiations with mandates to impose a plan to protect the earth's climate. Why not in reality?

The political systems of the most powerful countries are dominated by fossil fuel interests that want to go on emitting GHGs. They are supported by institutions, corporations, and constituencies that fear the consequences of a transition to a fossil free world. Many national governments suffer a "democracy deficit" that often makes conventional electoral politics and lobbying appear fruitless for ordinary people. National governments fear global climate protection may interfere with their pursuit of wealth and power. The dynamics of capitalism make climate protection policies appear a threat to prosperity. The world's dominant economic ideology, all-power-to-the-market neoliberalism, condemns anything that might interfere with the pursuit of private profit. And the institutions that supposedly represent the world's people, notably the UN, are in fact dominated by national governments and those who control them. Call it the world order of climate destruction.

This world order has so far proven to be insurmountable for climate protection strategies that operate exclusively within the framework of conventional electoral politics and lobbying. But finding ways to act effectively when conventional representative institutions fail is what social movements do. From the abolitionists to the civil rights movement to the Polish Solidarity movement to the Keystone XL pipeline blockades, when democratic channels have been blocked social movements have used "people power" direct action to do what people needed. Why have they been able to do so?

Gandhi once wrote, "Even the most powerful cannot rule without the cooperation of the ruled." The powers that are responsible for climate change could not continue for a day without the acquiescence of those whose lives and future they are destroying. They are only able to continue their destructive course because others enable or acquiesce in it. It is the activity of people—going to work, paying taxes, buying products, obeying government officials, staying off private property—that continually recreates the power of the powerful. A movement can impose its will without weapons or violence if it can withdraw that cooperation from the powers that be. Fear of such withdrawal can motivate those in positions of power to change.

The power of the people to impose their will on the powerful has been demonstrated repeatedly throughout history. The workers on the pyramids in ancient Egypt struck to force payment of their wages. The Indian civil disobedience campaigns won freedom from Great Britain. Unions and strikes around the world established labor rights for people who were little more than wage slaves. The movement for nuclear disarmament played a significant role in reducing the number of strategic nuclear weapons by 80%. The civil rights movement ended legal racial segregation in the American South. The feminist movement changed gender relations and expanded women's rights in multiple spheres of life in nearly every country in the world. The gay rights movement eliminated legal discrimination against gays in many countries and is now winning the right of gay people to marry. "People power" revolutions have overthrown oppressive regimes from Poland to the Philippines to Tunisia. These movements have certainly not created utopias, or even permanently realized all their goals, but they demonstrate that people in fact have the power to transform even brutal, violent, and oppressive systems and structures.

Of course, a collection of frightened, isolated, confused individuals will find it difficult to engage in such concerted action. So in order for "people power" to express itself effectively, people must organize themselves, gain the conviction that their action is necessary and right, and discover their power in action. That requires a social process that joins people together in a social movement, clarifies common interests, exposes the false arguments of the opposition, establishes a claim to moral and legal legitimacy, and engages in actions that reveal the potential power of the people.

Gene Sharp's monumental three-volume study *The Politics of Nonviolent Action* lays out 198 methods that have been used in such struggles, ranging from sit-ins to general strikes and from boycotts to dual sovereignty and parallel governments. He writes that if power is to be controlled by withdrawing help and obedience, "noncooperation and disobedience must be widespread and must be maintained in the face of repression aimed at forcing a resumption of submission." But once there has been a major reduction in the people's fear and a willingness to suffer sanctions as the price of change, "large-scale disobedience and noncooperation become possible." The ruler's will is thwarted "in proportion to the number of disobedient subjects and the degree of his dependence upon them."[1]

Nonviolent climate insurgency is an implementation of this approach. It provides a vehicle for the people of the world to organize ourselves to protect and restore the earth's climate system.

The immediate goal of the climate insurgency is the same as that enunciated by the nations of the world in the Paris Agreement: to prevent more than 1.5°C additional warming. That requires reducing atmospheric GHGs to 350 ppm or less. That in turn requires reducing developed country GHG emissions to essentially zero by 2050, plus the preservation and expansion of forests, farms, and other "sinks" that withdraw carbon from the atmosphere. And that in turn requires replacing all fossil fuel energy by renewable energy and energy efficiency by 2050. Many other non-climate goals can be combined with these, but these are the core goals for the climate insurgency.

The fundamental strategy for the nonviolent constitutional insurgency is to withdraw the support of the people from climate destruction. It uses nonviolent direct action—aka civil disobedience—to express popular refusal to acquiesce in the burning of fossil fuels and to force a transition to climate-safe energy. It defends such action as both the right and the duty of the people—and proclaims climate destruction to be illegal and unconstitutional. It mobilizes both those who are willing to engage in activities the authorities claim to be illegal and the wider population who support their objectives. It seeks to create an irresistible momentum of escalating popular action for climate protection.

The insurgency creates "inconvenience" for the fossil fuel forces and the governments that authorize their climate destruction so that they will decide they will be better off by shifting to a climate-safe path. But its ultimate "weapon" is to create a "dual power" that challenges the very legitimacy of businesses that threaten the climate and governments that permit or encourage them to do so.

PART II
CLIMATE
INSURGENCY
IN AMERICA

Chapter 4
A Strategic Vision

THE FUNDAMENTAL GOAL OF THE CLIMATE INSURGENCY IS not triumph or domination. It is a solution to the problem of global warming. That requires halting greenhouse gas emissions and replacing fossil fuels with renewable energy and energy efficiency.

The fundamental goal of the fossil fuel industry is to make a profit today and indefinitely into the future. It is supported by a wide range of other forces whose goals range from making money from the use of fossil fuel energy to strengthening their political power to keeping their access to heat, light, and transportation to keeping their jobs.

The fundamental goal of the climate insurgency will be met when, with or without the consent of the fossil fuel industry, fossil fuel burning has been halted and replaced.

A Strategic Assessment

The struggle between the climate insurgency and the supporters of fossil fuel involves what Gandhi called a "matching of forces." But the strengths and weaknesses of the climate destroyers and the climate protectors are anything but symmetrical.[1]

The fossil fuel industries possess immense wealth. Their coal, oil, and gas reserves in the ground are worth many trillion dollars; their financial assets are colossal; and their plants, pipelines, tankers, and other facilities span the globe. They control fossil fuel technology. The peoples of the world are dependent on them for the energy that fuels their vehicles and lights their houses. They wield immense political power through their wealth and the economic dependence of political jurisdictions. Their political allies include much of business and labor. They have a wide array of means to sabotage fossil free alternatives. They are experienced, savvy, and proactive about politics and shaping public opinion. Their freedom to do what they want with their property is supported by law, government, and much of public opinion.

But the fossil fuel forces also possess enormous if asymmetrical weaknesses. Their greatest weakness is that their business plan is climate doom. Their continued operation threatens the life and liberty of every individual on earth, not to mention our posterity. They are also to blame for massive damage in many localities. Despite vast public relations spending they are widely despised not only for their environmental but also for their economic and political impact. Their business and labor allies have fundamentally different interests and would rapidly abandon them if they found it advantageous to do so. The dependence on fossil fuels that forms the core of their power can be eliminated by renewable energy and energy efficiency.

The climate protection forces represent the common interest of the world's people. They are essential to protecting the life and liberty of every person on earth. They embody the moral imperative to halt global warming. Their success is essential to the well-being of our posterity. And whatever the courts may say, they have the right and duty to prevent climate doom.

The climate protection forces also have significant weaknesses that have so far prevented them from achieving their goals. Most people feel isolated and therefore powerless in the face of the fossil fuel establishment. They feel that without fossil fuels they are likely to freeze to death in the dark. Many people are poorly informed about the causes of climate change and what is necessary to halt it. They believe themselves powerless to affect it. The climate protection movement is largely isolated from the great majority of people whose interest it represents. It is structurally disempowered by national and global "democracy deficits" that allow private interests to block the formation and implementation of the popular will. It lacks an effective strategy for achieving its objectives.

A Strategy for Climate Insurgency

To realize their objectives, the climate protection movement needs to weld the people of the world into an effective force capable of compelling corporations, governments, and institutions to shift from fossil fuels to clean energy. That requires creating understanding and determination in every community, geographical and virtual. It requires breaking down the invisible rules that currently discourage people from thinking and talking about the one thing that will most determine their future and that of their posterity. It means self-organization through which people move out of isolation and become part of a movement. It means creating ways for people to reach out and join with others. Conversely, it also means creating ways that those who are already organized can reach out and draw in those who are not. This underlying process of de-isolation and movement construction is the necessary condition for the ultimate success of the climate insurgency.

That process both depends on and contributes to a transformation of the public mind. People need a clear understanding that continuing to burn fossil fuels will destroy the climate system. They need to understand how fossil fuels can be rapidly replaced by renewable energy and energy efficiency. And they need to understand the power of the people to impose such a change. The strategy of the climate insurgency is designed to bring about that transformation of the public mind.

The climate insurgency will make two basic demands on every business, government, and institution. The first is to immediately halt all new fossil fuel infrastructure—a "fossil freeze." The second is to create and immediately start to implement a Climate Action Plan that lays out a credible pathway for the elimination of fossil fuel use by 2050.

The climate insurgency will develop the power to impose these changes by:

Taking actions that change millions of people. The most important targets of the climate insurgency are the hearts and minds of our friends and neighbors locally, globally, and virtually. The goal is to encourage them to organize themselves and act to protect the climate.

Undermining fossil fuel's "pillars of support." The climate insurgency will inspire people to force the institutions in which they live to withdraw support from the fossil fuel agenda. They will fight within churches, unions, governments, and other organizations of all kinds to end complicity with the fossil fuel industry and to develop and implement plans to go fossil free.

Reducing dependence on fossil fuels by implementing alternatives. The greatest power of the fossil fuel forces is the fear that without their product we will all shiver to death in the dark while we watch jobs disappear and the economy collapse. The climate insurgency will counter this fear by proposing actionable plans for rapidly replacing fossil fuels with renewable energy and energy efficiency and forcing governments, businesses, and other institutions to implement them.

Discrediting the legitimacy of the climate destroyers. The climate insurgency will define the fossil fuel forces as criminals laying waste to the common heritage of humanity. It will challenge their right to use the property they claim to own and the authority they claim to hold as long as they use it for climate destruction. It will assert that it is the climate protectors who represent legitimate law and authority. It will make these claims on the basis of fundamental human and constitutional rights that require governments to protect the earth's climate system as part of the public trust. It will call on all good citizens to exercise their right and duty to protect that public trust.

Increasing the negative consequences of continuing fossil fuel extraction and burning. The climate insurgency will directly "inconvenience" the fossil fuel industry and the governments that permit its climate destruction by occupations, blockades, mass picketing, and a wide range of other direct action tactics. It will challenge the process by which governments give private corporations permission to destroy the climate. It will turn the public against the fossil fuel industry in a way that threatens its future functioning and profitability. It will thereby force fossil fuel companies and their investors to retreat from their plans for continuing and expanding their fossil fuel operations.

Developing a "dual power." If the forces of climate destruction do not voluntarily reverse their course in the face of popular and insurgent pressure, the insurgency will encourage the people to withdraw their support and cooperation from the established authorities and form their own tribunals and popular assemblies to authorize the fossil freeze and impose Climate Action Plans. The threat to established authority posed by the development of dual sovereignty and parallel government will provide the ultimate sanction against those governments that continue to authorize climate destruction.

Integrating a wide range of other popular needs and concerns into plans for climate protection. Human needs and concerns go far beyond climate protection. The climate insurgency will demand that Climate Action Plans provide protection for those they may adversely affect and greater justice for all.

Creating a global climate protection race. The insurgency will instigate a competitive bidding war among nations, businesses, and institutions to demonstrate their commitment to climate protection. It will help construct a "coalition of the willing"—those who are willing to halt all new fossil fuel infrastructure and implement adequate Climate Action Plans. And it will instigate governmental and popular sanctions against those who don't.

These elements are mutually reinforcing, and they all need to happen all the time. Every climate insurgency action should be designed to strengthen one or several of these means. They will be successful if they draw people out of isolation and despair into action; increase the number, unity, and power of participants in the movement; build understanding and support in the wider population; weaken the fossil fuel forces; undermine their pillars of support; give climate destroyers motivation to change; threaten loss of their legitimacy and authority if they don't change; and lay the groundwork for further action. As Bill Moyer wrote, "Social movements involve a long-term struggle between the movement and the powerholders for the hearts, minds, and support of the majority of the population."[2]

Chapter 5
Imposing a Fossil Freeze

IN EARLY 2016, TWENTY-ONE YOUTH PLAINTIFFS SUED THE United States government in federal court demanding it meet its legal duties to protect their rights to life, liberty, and the public trust from the growing harm climate change was causing to them and their posterity. Perhaps some day the legal system will enforce those duties—but in the meantime, we the people and our posterity are being subject to irreversible harm. The arguments brought by the "climate kids" also provide a justification for the rest of us to exercise our right—indeed our duty—to take the law into our own hands. That is what the climate insurgency is doing.

The climate kids asked the court to do two things: first, order the U.S. government to halt construction of new fossil fuel infrastructure; second, institute a Climate Action Plan to halt fossil fuel emissions before it is too late. This chapter will explore how we the people can impose a halt to new fossil fuel infrastructure. The next chapter will examine how we can force governments, businesses, and institutions to institute and implement Climate Action Plans to end fossil fuel emissions entirely.

Without pipelines, oil trains, power plants, and other infrastructure, coal, oil, and gas would remain harmlessly in the earth where they have lain for millions of years. All over the world, campaigns like Break Free from Fossil Fuels have focused on such fossil fuel infrastructure. The climate insurgency will integrate such actions through a demand for no new fossil fuel infrastructure—for a "fossil freeze."[1]

Fossil fuel projects are multiplying—and many are meeting effective resistance both through legal means and through civil disobedience. Last year, after a seven-year struggle, President Obama turned down a permit for the Keystone XL pipeline. Collateral struggles have terminated other Canadian tar sands projects worth $17 billion. Scotland, Wales, France, and Tasmania, as well as states like New York, have banned natural gas fracking. On the West Coast, four of the six proposed giant coal ports have been cancelled. In southern India, a six-year campaign has stopped a huge coal plant.[2] Portland, Oregon, has become the first city in the U.S. to pass a resolution opposing the development of all new infrastructure for fossil fuel transport and storage.[3]

New fossil fuel infrastructure projects have an operational and economic life span of more than thirty years.[4] If the average lifetime of existing fossil fuel infrastructure is less than thirty years, then a freeze on new fossil fuel infrastructure will largely eliminate fossil fuel burning by 2050. Even without other policy changes, a freeze on new fossil fuel infrastructure will realize the IPCC targets as existing fossil fuel facilities are retired at the end of their useful lives. It will provide little threat and much benefit to existing workers and communities as long as the foregone fossil fuel investment is applied to renewable energy and energy efficiency.

State and local policy is increasingly rejecting new fossil fuel infrastructure based on the greater economy of grid modernization, distributed energy, energy efficiency, and the falling cost of renewables. In many localities advocates of that change could be tacit or explicit allies of a fossil freeze campaign.

Of course, there is a commonsense objection to proposals to "leave fossil fuels in the ground": Won't billions of people soon freeze to death in the dark? However, a freeze on new fossil fuel *infrastructure* does not pose such a threat. Shifting *new investment* from fossil fuels to clean energy provides a commonsense answer: Why should we spend one penny for costly, outmoded, environment-destroying fossil fuel infrastructure when new clean energy technology is not only climate-friendly but cheaper? It recognizes that we can't go completely "cold turkey" with our fossil fuel addiction. At the same time, it provides a crucial part of the pathway to a fossil free economy.

Public opinion often supports an "all the above" energy policy, including new fossil fuel infrastructure. The fossil freeze takes this crucial problem head on. It provides an opportunity to educate the public on the core issue: abolishing the use of fossil fuels. And it forces people and institutions to decide whether they are serious about fighting global warming: Are they willing to walk the walk as well as talk the talk?

Fossil fuel infrastructure projects depend on a long development process with many vulnerable points where the climate insurgency can intervene:

Infrastructure planning. The energy system is highly integrated; it requires and is the result of long and detailed planning by federal and state governments and energy companies. This process depends on agreement by public regulators. It is often subject to legislation. Every planning decision that perpetuates fossil fuel use rather than replacing it by clean energy is a potential point at which to demand a fossil freeze.

Permitting. Every infrastructure project requires permits from public authorities. The permitting process provides many legal requirements and many opportunities for public intervention. While regulators and companies often collude to evade such responsibilities, such collusion in turn provides targets for further intervention. In particular, the fossil freeze forces should insist that no permits for fossil fuel projects ever be given without an honest comparison to the costs and benefits of alternative clean sources of energy.

Financing. Fossil fuel projects are expensive. They generally depend on both public and private financing as well as on public policy. The private banks, financial institutions, and investors who finance fossil fuel projects—and profit from them—are important targets. So are government agencies that float or guarantee financing for projects. Targeting of financing can make investors fear the risks such investments carry and make them wary of the cost of fossil fuel investments to their reputations.

Construction. The building of fossil fuel infrastructure has been the target of dramatic direct actions that have delayed projects and educated the public about their dangers. The Keystone Blockades and the actions against the Kinder Morgan gas pipeline in Massachusetts and New Hampshire played a significant role in their loss of support and ultimate defeat.

The climate insurgency can call for the halting of all new fossil fuel infrastructure projects. It can demonstrate its opposition and offer alternatives wherever such projects are proposed. But nonviolent movements, as Gene Sharp wrote, need to concentrate action on "the weakest points in the opponent's case, policy, or system" to strengthen the relative power of the movement. Targets should be chosen that symbolize the broader "evil," which are least defensible, and which are capable of arousing the greatest strength against them.[5] The insurgency should concentrate its attentions on carefully chosen targets that meet as many as possible of these criteria:

Threat to climate. The national campaign against the Keystone XL pipeline featured James Hansen's statement that if tar sands oil were burned, it was "game over" for the climate.

Threat to local people and environment. Fossil fuel projects, in addition to their climate effects, almost always have local side effects that greatly strengthen the opposition to their extraction, transportation, and burning. Native people and ranchers—the self-described "Cowboy and Indian Coalition"—played a major role in the campaign against the Keystone XL pipeline. People in Albany's South End played a critical part in the campaign against their threatening neighbors the "bomb trains."

Local support. Do such factors, and others, lead to broad local support for actions? Conversely, are there factors—such as strong community, trade union, or business support for a project—that make a campaign more difficult?

Allies. Allies can play an important role in supporting a struggle. Are there religious communities that have taken a strong stand on climate? Are there unions that strongly support clean energy alternatives?

Cost. Many projects are costly boondoggles that enrich the builders but are unnecessary or even a financial burden for local and regional communities and energy consumers.

Availability of alternatives. The strongest public argument for a fossil fuel project is usually the need for energy. The availability of clean energy alternatives is often an effective argument against new projects. The proposed Smith power plant in eastern Kentucky, for example, was defeated in large part because opponents were able to show that clean energy alternatives could meet energy needs at a lower price and with substantial human and environmental benefits for the local region.[6]

Injustice. Fossil fuel projects are notoriously concentrated in communities of color and poverty. They disproportionately threaten young people and future generations, since they are the ones who will have to suffer the consequences, ranging from disease to climate destruction. Projects that exemplify such injustice provide an additional justification for opposition.

Vulnerability of identifiable decision-makers to pressure. When Bill McKibben launched the national campaign against the Keystone XL pipeline, he emphasized the fact that Barack Obama had the personal authority as president to veto its permit; unlike many other decisions, this one could not be overridden by Congress. On that basis he targeted the White House for sit-ins, and throughout the campaign activists kept the focus on the president's decision.

Moral responsibilities. Institutions that claim to embody moral responsibility and leadership, such as churches and universities, are particularly vulnerable to pressure to meet high ethical standards. It is no wonder that they have often been targeted by campaigns to divest from fossil fuels, to invest in clean energy, to reduce their own use of fossil fuel energy, and to develop green alternatives for themselves and the communities around them.

Democratic accountability. Private companies and institutions may claim the right to do what they wish, but government officials purport to follow the law and elected officials claim to be accountable to the people. When such officials align themselves with the private interests of fossil fuel companies, they may be particularly vulnerable to movement action.

Public disapprobation. Many corporations, such as oil, gas, and coal companies and public utilities are already despised by much of the public for long trails of abuse to consumers, the environment, and the public. Such companies are likely to make good targets because they are already distrusted and few tears will be wasted over their distress.

Vulnerability to exposure. Corporations and political leaders have often swathed fossil fuel projects in lies, manipulation, corruption, and abuse of democratic government.

Such forces are often particularly vulnerable to disclosures that damage their reputations.

Hypocrisy. Companies, governments, and officials regularly make claims and commitments to appear as "good guys" to the public. They may claim that they are operating in accordance with legal safety requirements or promise to reduce their greenhouse gas emissions. Such claims offer two "handles." First, they appear to accept the movement's premise: why, for example, should a company promise to reduce greenhouse gases unless it accepts the basic premise that they are harmful and need to be reduced? Second, when they don't live up to their commitments, they are guilty not only of the harm they are doing, but of the hypocrisy of lying about it.

Vulnerability to mass nonviolent direct action. Some projects are difficult to target directly; while intrepid Greenpeace actions have sometimes targeted offshore drilling rigs, for example, they are probably not good direct action targets for community groups. Company headquarters, government offices, or investors may make better targets in such situations. Some targets, conversely, are particularly propitious: the White House was an excellent target for the Keystone XL sit-ins because they could focus attention on the president; easily attract the media; involve a police force that is experienced at dealing with protesters; and deter repression and brutality to protesters because of its visibility and symbolic role.

While there are no perfect targets, ones that combine several of these features will be particularly advantageous.

Framing of the message can take advantage of the strengths of the movement and the weaknesses of those who advocate new fossil fuel infrastructure. For example, every campaign against a fossil fuel project can point out how the project contributes to climate destruction; how it threatens local communities and environments; and how it aggravates racial, class, gender, geographical, and generational injustice. Campaigns can emphasize the support of the local community and allies. They can explain the financial burden a project puts on local communities and propose clean energy alternatives. They can identify decision-makers and hold them accountable; point out the duty of elected and appointed officials to defend the interests of their constituencies; and raise up the ethical responsibility of religious organizations, universities, and other institutions that claim to provide moral leadership. They can note public disapproval of the fossil fuel companies, expose their lies and manipulations, and demand that they live up to their commitments.

The purpose of the fossil freeze campaign is to halt the building of new fossil fuel facilities—and to show the public the necessity, the rightness, and the power of the people taking action together, as exemplified in the actions of the climate insurgency.

Chapter 6
Imposing Climate Action Plans

CLIMATE PROTECTION REQUIRES NOT ONLY A HALT TO NEW fossil fuel infrastructure but also a planned transition to a fossil free society. That means not just changing energy sources but changing every institution that uses energy. And it means using forests, farms, and other "sinks" to begin withdrawing GHGs from the atmosphere.

This transition requires planning. The climate insurgency will insist that every institution—from schools, churches, and municipalities to the federal government—develop and immediately start to implement a Climate Action Plan that will eliminate fossil fuel burning by 2050 at the latest.

Many governments, corporations, and institutions have established Climate Action Plans that lay out their pathway to reducing greenhouse gases. However, many of these reductions are too little and too late, many plans are not mandatory, many are not being implemented, and many organizations have no Climate Action Plan at all. In short, these institutions are in default on their obligations to protect the public trust.

The suits brought by youth plaintiffs to protect their public trust and other constitutional rights against climate destruction call for enforceable remedial plans to phase out fossil fuel emissions and draw down GHGs to stabilize the climate system.[1] They lay out the parameters that such plans must realize:

> Global atmospheric CO_2 concentrations must be reduced to below 350 ppm by the end of the century.

That requires a near-term peak in CO_2 emissions and a global reduction in CO_2 emissions of at least 6% per year, alongside approximately 100 gigatons of carbon drawdown this century from global reforestation and improved agriculture. If significant annual emission reductions are delayed until 2020, a 15% per year reduction rate will be required to reach 350 ppm by 2100. If such reductions are delayed beyond 2020, it may not be possible to return to 350 ppm until 2500 or beyond.[2]

In April 2016, King County Superior Court Judge Hollis Hill, in response to a suit by Washington climate kids, ordered the State of Washington's Department of Ecology to promulgate an emissions reduction rule by the end of 2016 and make recommendations to the state legislature on science-based greenhouse gas reductions in the 2017 legislative session. Andrea Rodgers, the climate kids' attorney, said, "For the first time, a U.S. court not only recognized the extraordinary harms young people are facing due to climate change, but ordered an agency to do something about it." The Department of Ecology is now "court-ordered to issue a rule that fulfills its constitutional and public trust duty to ensure Washington does its part to reduce greenhouse gas emissions and protect the planet."

Judge Hill noted the extraordinary circumstances of the climate crisis, saying, "This is an urgent situation. . . . These kids can't wait." Neither can the climate insurgency. The climate insurgency, representing the right of the people to climate self-protection, will impose similar requirements on every political jurisdiction, business, and institution to develop Climate Action Plans.

It is not the role of the climate insurgency to determine in detail the specific pathway institutions will take to eliminate fossil fuels—that is a task for many different groups, inside and outside the insurgency, to determine. The role of the climate insurgency is to make clear the negative consequences—from climate change and from movement action—that will befall those who do not take such measures.

Further, the climate insurgency can, like Judge Hill, insist that organizations adopt Climate Action Plans that include targets and timetables to reduce greenhouse gas emissions at the pace required by climate science—eliminating them by 2050 or sooner. Plans must include intermediate goals that do not leave the heavy lifting for the last years of the transition.[3] They must begin to cut GHG emissions immediately—and the insurgency must be prepared to challenge them if they do not.

While the insurgency will ask every institution to develop and implement a Climate Action Plan that meets these requirements, it will concentrate its campaigns on carefully selected targets that meet criteria like those laid out in the previous chapter for infrastructure campaigns. For example, it will select governments, businesses, and institutions that emit large amounts of greenhouse gases; threaten local communities and environments; and aggravate race, class, gender, and geographic inequalities. It will focus in particular on targets that are subject to pressure from internal constituencies and the broader public.

The climate insurgency will need to develop means to judge the adequacy of Climate Action Plans. This could start with a simple rating system or scorecard. It could also involve some kind of expert panels. The role of people's climate tribunals in legitimating the demand for Climate Action Plans and evaluating their adequacy is discussed in the chapter "Dual Power" below.

Of course the most critical Climate Action Plan will be that of the U.S. government. It presented a Climate Action

Plan of sorts as a pledge to the Paris climate conference—but a totally inadequate one. Its pledge was to drop GHG emissions 26% below 2005 levels by 2025. According to an analysis by Climate Interactive and MIT Sloan, that, along with the pledges of other countries, will lead to a global temperature increase of a catastrophic 3.5°C (6.3°F) above preindustrial levels.[4]

That analysis finds that to reduce warming to 1.8°C (3.2°F) will require the U.S. to increase its pledged GHG reduction from 26% to 45%, and for other countries to make comparable emission reductions. And of course it will require the U.S. to essentially eliminate GHG emissions by 2050.[5]

The climate protection movement has consistently advocated a shift from fossil fuels to renewable energy and energy efficiency. But it has had difficulty translating that advocacy into concrete demands that decision-makers can be pressured to implement. The demand to establish and implement Climate Action Plans that will eliminate fossil fuels by 2050 may provide a way to overcome that difficulty.

Chapter 7
Self-Organization for Climate Defense

POWER GROWS FROM THE ABILITY TO COOPERATE.[1] THE people of the world have not halted global warming because we are disorganized. The climate insurgency is first and foremost a means to help people organize themselves and coordinate their action.

Chapter 1 of this book began with a small meeting in a church in Albany, New York. People from the local community and diverse others from the city and the region came together around common concerns—both the direct threat oil trains posed to the neighborhood and the threat that burning fossil fuels posed to all the people of the world. They organized themselves to act. Such self-organization is what can halt global warming.

The climate insurgency is not "an organization" with membership, dues, and the rest. It is a network that includes people who are members of many organizations and it coordinates with many organizations. People "join" the climate insurgency by meeting with other people and acting together with them, virtually and in the flesh.

The Break Free from Fossil Fuels days of action showed that those resisting fossil fuels are able to coordinate their action locally, regionally, nationally, and globally. Scores of organizations from every part of the globe defined common goals: an end to new fossil fuel infrastructure and a rapid transition to renewable energy and energy efficiency. They agreed to mass mobilization and nonviolent direct action as the means. They identified a specific set of sites where action would be concentrated. They set a timetable that would allow them to magnify the effect of local actions around the world. And they established a network of fossil free forces that will allow other actions to be similarly coordinated in the future.

Self-organization begins in the milieus in which people live their lives—the workplace, the neighborhood, and the many physical and virtual communities of which people are part. Currently an invisible wall of silence seems to block the sharing of concerns about global warming within these milieus. According to one recent poll, only one in four Americans say they "often" or "occasionally" discuss global warming with family or friends.[2] Three in four Americans say they talk about global warming "rarely" or "never."[3] A tacit rule of silence forms an initial barrier to organizing and acting to protect the climate.

When people transgress their milieus' unwritten rules against talking about climate change, they are making a first step toward climate insurgency. They are joining with others in their own milieus to learn and to share their feelings and ideas. That can create a new basis for action. (In the next two chapters we will discuss how the climate insurgency can help stimulate people to break down those barriers.)

The next step is for those who have become concerned to reach out to others beyond their milieus. They need to connect with other communities like their own. And they need to connect with networks and organizations that are already acting for climate protection. For that to happen those networks need to provide easy and congenial channels through which new people can link up and begin to participate in the wider movement. As Mao Zedong said of the guerrilla, the climate insurgent must "move amongst the people as a fish swims in the sea."

These groups of diverse origin need to network with each other to form common goals and coordinate their plans and actions. While many different organizations may participate in this process, the movement itself needs to transcend particular organizations and overcome divisions into silos and turf. Actions like Break Free have shown that it can be done.

The climate insurgency needs large numbers of people actively involved, not only marching and getting arrested, but doing research, leafleting, holding community meetings, serving on committees, and using social media. The prevailing pattern of the climate insurgency has been to form temporary leadership groups for particular actions and campaigns. Individuals take on leadership roles in accordance with their interests and abilities and the needs of the moment.

This process of breaking down the barriers within milieus and connecting with others beyond them changes the social relations that have made people powerless to halt climate change.

Chapter 8

Turning the Public against Fossil Fuels

MORE THAN A QUARTER-CENTURY AGO, CLIMATE SCIENtists established that global warming was happening, that it was caused by burning fossil fuels, and that continuing to burn them would have catastrophic consequences. They also made clear that halting that burning would prevent the worst effects of global warming, and that a transition to clean energy could be accomplished without catastrophic consequences. Today, 97% of climate scientists agree that human-made global warming is happening.

Climate scientists have persistently reported these findings to government authorities and the public. However, they have reached the public through filters that have distorted public understanding. A persistent campaign, financed by the fossil fuel industry, has sought to sow public doubt about the reality of global warming and the need to do anything about it. Major media have falsely presented human responsibility for global warming not as established scientific fact but as something on which a significant proportion of climate scientists disagree. A small but vocal and well-funded right-wing political sector has proclaimed climate change a hoax.

Most Americans accept the reality of climate change, but they see it through a haze of uncertainty. Their views, however, are evolving. The observable effects of climate change, from extreme weather events to rising sea levels, allow people to draw their own conclusions from visible evidence. The earth itself is weighing in on the "climate debate."

Effective climate protection requires that the public be convinced of three things:[1]

> 1. Elimination of fossil fuels is necessary to limit climate catastrophe.
> 2. Fossil fuels can be eliminated without causing other catastrophes.
> 3. People, acting together, have the power to eliminate fossil fuels.

This chapter addresses how the climate insurgency can help change the public understanding of the threat of climate change and what to do about it. The next chapter addresses how to overcome the sense of powerlessness that leads to passivity. Both chapters make cautious use of public opinion polling data to help understand the public mind.[2]

Climate Change in the Public Mind

Three premises are required for the public to believe that the elimination of fossil fuels is necessary: climate change is real; it is caused by burning fossil fuels; and continuing to burn them will have catastrophic consequences.

Climate change is real. Most Americans recognize the reality of climate change. In one recent poll: "Two in three Americans (68%) think global warming is happening. Large majorities of Democrats (86%)—liberal (92%) and moderate/conservative (79%)—think it is happening, as do two in three Independents (68%, up 9 points since Spring 2014) and liberal and moderate Republicans (65%)."[3]

Climate change deniers are a small proportion of the population. Those described as "dismissive," who are "certain that global warming is not occurring, tend to regard the issue as a hoax and are strongly opposed to action to reduce the threat" are only 15% of the population.[4]

Climate change is caused by burning fossil fuels. Most Americans think humans cause climate change. For example: "Just over half of registered voters (52%) think that global warming is caused mostly by human activities, with an additional 6% saying human activities and natural changes both play a role. A large majority of Democrats (72%, and 82% of liberal Democrats), four in ten liberal and moderate Republicans (43%), but only two in ten (22%) of conservative Republicans think global warming is mostly human-caused."[5] (Like most polls, this one does not make clear what proportion of those who believe climate change is caused by human activities understand that it is overwhelmingly caused by burning fossil fuels.)

Continuing to burn fossil fuels will have catastrophic consequences. In a recent poll, 75% of Americans said that global warming was already having a serious environmental impact or would in the future. Nine in ten Democrats agreed, compared with 58% of Republicans. One-third of Republicans said they believed it would never have much of an impact on the environment.

Most Americans are worried about climate change. "Over half of Americans (56%) say they are 'very' or 'somewhat' worried about global warming. Liberal Democrats are the most worried (83%), followed by moderate/conservative Democrats (66%). About half of Independents (53%) and liberal/moderate Republicans (50%) are worried about global warming. Relatively few conservative Republicans (28%) are worried," though the number has been increasing.[6]

There is division within the conservative Republican camp. For example, "only about one in three conservative Republicans (36%) thought reaching an agreement" on global warming in Paris was important.[7] And "37% of conservative Republicans support setting strict limits on carbon dioxide emissions."[8]

Two additional premises are necessary for the public to support action: eliminating fossil fuel burning will reduce climate change; fossil fuel burning can be eliminated without catastrophic consequences.

Climate destruction can be reduced without devastating consequences. Most Americans believe that their country can help reduce global warming, and that its doing so would be beneficial. In one recent poll, half or more of registered American voters said that if the United States takes steps to reduce global warming, it will provide a better life for our children and grandchildren (64%), improve people's health (59%), and save many plant and animal species from extinction (55%).[9]

Most Democrats, particularly liberal Democrats, expect these and other benefits, as do at least half of Independents and liberal/moderate Republicans. However, fewer than half of conservative Republicans expect any of these benefits if the U.S. takes steps to reduce global warming.[10]

Many people expect deleterious but not disastrous side effects of climate protection. "Over half of registered voters think that if the United States takes steps to reduce global warming, it will cause energy prices to rise (57%). Relatively few Americans think it will cost jobs and harm our economy (28%) or interfere with the free market (27%). Republicans, particularly conservative Republicans, are the most likely to expect these negative consequences."[11]

Public Opinion on Climate Policy

Americans support clean energy. Of Americans registered to vote, 84% support more research into renewable energy sources such as solar and wind power; 80% support tax rebates for energy-efficient vehicles or solar panels; 74% support regulating carbon dioxide as a pollutant; and 66%, including more than half of Republicans, support requiring fossil fuel companies to pay a carbon tax and using the money to reduce income and other taxes by an equal amount.[12]

"Prior to the start of the 2015 [Paris climate] conference six in ten Americans (62%) said that it was important that the world reach an agreement in Paris to limit global warming. This opinion was most widely held by liberal Democrats—nearly 9 in 10 (87%) said it was important to reach a climate agreement in Paris. At least six in ten moderate/conservative Democrats (68%), Independents (62%), and liberal/moderate Republicans (65%) also thought it was important to reach an agreement. By contrast, only about one in three conservative Republicans (36%) thought reaching an agreement . . . was important."[13]

"Two in three registered voters (65%) support setting strict carbon dioxide emission limits on existing coal-fired power plants to reduce global warming and improve public health, even with the explicit caveat that the cost of electricity to consumers and companies would likely increase. Most likely to support the limits are Democrats (84%, 92% of liberal Democrats) as well as Independents (65%, up 17 points since Spring 2014) and liberal/moderate Republicans (66%). However, only 37% of conservative Republicans support setting strict limits on carbon dioxide emissions."[14]

Americans generally favor regulating business activity more than taxing consumers. There is broad support for capping power plant emissions. Half of all Americans and 58% of Democrats said they thought the government should take steps to restrict drilling, logging, and mining on public lands, compared with 45% who opposed such restrictions. But just one in five Americans favored increasing taxes on electricity as a way to fight global warming; six in ten were strongly opposed, including 49% of Democrats. And support was not much higher for increasing gasoline taxes, at 36% overall.[15]

Despite the fact that over half of Americans are worried about climate change and support clean energy, most still support an "all of the above" energy policy that maintains or even increases burning of fossil fuels, even in such high-risk forms of extraction as offshore drilling. For example, 60% of Americans registered to vote support the expansion of offshore drilling for oil and natural gas off the U.S. coast, including 62% of Independents, 79% of Republicans, and 42% of Democrats.[16]

How the Climate Insurgency Can Change Public Views

There will never be 100% agreement on anything, even whether Elvis is alive. However, the climate insurgency, combined with the effect of other forces, can shift public opinion on the crucial point: climate change is not just caused by some vague "human activity," but by the very specific activity of burning fossil fuels; therefore it is essential not just to increase clean energy, but to end the burning of fossil fuels.

Effective climate action requires not just belief that climate change is real, or even that it is human-caused, but an understanding that it is specifically caused by fossil fuel emissions. A central goal for the climate insurgency must be to move the public to the conviction that the greenhouse gases released by the burning of fossil fuels are the principal cause of climate change. It needs to persuade the majority of Americans who are worried about climate change that the elimination of fossil fuels is not only desirable but necessary.

In a recent study investigating the degree of scientific consensus on climate change, researchers determined that

"97% of climate scientists are convinced human-caused global warming is happening." However, the average American registered voter estimates that fewer than six in ten climate scientists are convinced (59%).[17]

The distortions of scientific opinion promulgated by the fossil fuel industry, media, and conservative politicians can be challenged directly. For example, the Weather Channel has been notorious for concealing the connection between extreme weather events and global warming. The insurgency can use direct action tactics to demand that it and other media accurately report scientific opinion on climate questions. Similarly, the fossil fuel companies and their supporters can be exposed and isolated for their lies; campaigns against the climate denialism of the Heartland Institute and the U.S. Chamber of Commerce, for example, led important corporate supporters to quit them. The falsehoods promoted by conservative political officials can be targeted as out of step with the great majority even of Republicans, who do not share their views.

The insurgency needs to present an explanation of climate change that is scientifically informed but also simple, credible, and accessible to informed common sense—one that people can themselves explain to their friends and neighbors. The public essentially needs to understand that when rays from the sun hit the earth and are turned into heat, the emissions in the atmosphere from burning fossil fuels prevent that heat from escaping into space. The excess trapped heat is equivalent to the energy from four hundred thousand Hiroshima bombs a day. The result is total disruption of the climate system, leading to extreme results like heat waves and blizzards, droughts and floods, and other seemingly paradoxical outcomes. The solution is to stop putting fossil fuel emissions in the atmosphere so the sun's heat can again escape.

Fortunately, the climate insurgency has a powerful ally in highlighting climate change: the earth itself. The climate insurgency can call attention to impacts of climate change here and now. Every climate-related storm, flood, fire, and drought provides a teachable moment that can be made a public education curriculum of climate change. A prime example was Occupy Sandy Relief, initially a coalition of Occupy Wall Street and 350.org, which provided dramatic aid for the victims of Superstorm Sandy while pointing out the relation between Sandy's devastation and global warming. Action can also be taken at vulnerable sites—for example, tours and marches to mark increasingly vulnerable flood plains. The message to those who think climate change is only a problem in the distant future is that they can see with their own eyes that climate change is now.

For calling attention to the real cause of global warming there is no better means than dramatic action at fossil fuel facilities, corporations, and investors. The climate insurgency's campaign for a fossil freeze is not only a form of sanction against new fossil fuel infrastructure, it is a way to educate the public that burning fossil fuels is the dominant cause of climate change and that halting it is the essential condition for climate protection.

Changing public opinion on climate change is not so much a matter of changing individuals as of changing milieus. Personal persuasion by trusted friends, family, and neighbors is crucial. According to one study: "Only climate scientists (82%) are more trusted on global warming than interpersonal contacts (77%)."[18] The efficacy of the climate insurgency flows in large part from its ability to erode the invisible walls that prevent people from sharing their concerns about climate change and seeking the facts that media and special interests have distorted.

People are also responsive to messages from those they identify with. First responders to climate disasters have been important climate educators; nurses, for example, have come forward as particularly effective spokespeople on climate change. Mothers, doctors, firefighters, high school students, and community leaders all have a special role to play with their peers and communities. [19]

The climate insurgency also needs to isolate and deactivate the small but vocal minority who deny climate change and oppose climate action. The principal goal here is not to persuade them that they are wrong, but to erode their credibility in the rest of the population. This can involve factual refutation and exposure of hidden motives. It need not involve unfair attacks that lead others to defend or support them—fair ones should be sufficient.

Denial of human-created climate change is declining in the face of increasing scientific certainty and direct observation and experience. However, the next big barrier to climate action is the argument that global warming is now inevitable and that there is nothing people can do to halt it. The climate insurgency needs to persuade the public that society can forestall climate change by transition from fossil fuels. To do so it needs Climate Action Plans that show this is possible. It needs examples showing that it can be done and how. And it needs exemplary action that demonstrates that the people can defeat the climate destroyers.

Chapter 9
Turning Climate Worriers into Climate Warriors

WELL OVER ONE HUNDRED MILLION AMERICANS ARE WILL-ing to act to protect the climate. More than half of Americans say they are willing to sign a petition or make consumer purchases to reward or punish companies for their actions on global warming. Yet in fact most don't act. The main reason people don't act is probably that they doubt their action will be effective. Overcoming that doubt is a core objective of the climate insurgency.

Willing to Act . . .

According to polls, about half of Americans (53%) say they would "definitely" or "probably" sign a petition about global warming if asked by a person they like and respect. About four in ten say that, if asked, they would sign a pledge to vote only for political candidates that share their views on global warming (39%), attend a neighborhood meeting to discuss global warming and actions people can take (38%), or attend a public meeting or presentation about global warming (38%).[1]

More than one American in four says they have engaged in consumer activism in the past twelve months to reward or punish companies for their actions regarding global warming. Moreover, about half of all Americans say they intend to engage in this type of consumer activism in the next twelve months.[2]

Only about one American in ten (13%) has contacted a government official about global warming by letter, e-mail, or phone over the past twelve months; three in four of those who did (73%) urged them to take action to reduce global warming.[3] However, one in five intends to urge government officials to take action to reduce global warming over the next twelve months. Three in ten (29%) say they would be willing to join a campaign to "convince elected officials to take action to reduce global warming,"[4] while 36% of Americans say they have joined or are willing to join a campaign to convince elected officials to pass laws increasing energy efficiency and the use of renewable energy as a way to reduce America's dependence on fossil fuels.[5] That's more than seventy million people.

Perhaps most astonishing, one in four Americans (24%) would support an organization engaging in nonviolent civil disobedience against corporate or government activities that make global warming worse, and about one in six (17%) say they would personally engage in such activities.[6] That's forty million adults. Yet only a few thousand people have in fact committed civil disobedience to protect the climate.

. . . but What's the Use?

Despite their expressed willingness to act, most people don't. In one poll, even among the 16% of the population who are certain global warming is happening, understand that it is human-caused and harmful, and strongly support societal action to reduce the threat, only about a quarter have engaged in political activism on climate.[7]

A 1999 study based on poll data and focus groups noted that according to one survey "three out of four Americans" believe that "the earth's atmosphere is gradually warming as a result of air pollution and that, in the long run, this could have catastrophic consequences." Most, however, felt they were personally powerless to halt global warming, and that indeed the problem was insoluble. "People literally don't like to think or talk about the subject." Their concern "translates into frustration rather than support for action."[8]

A similar sense of powerlessness prevails today. A recent survey found that "relatively few respondents viewed any form of activism as effective." Donations were viewed as "highly" or "pretty" effective by 22% of the respondents; contacting government officials by 15%; and attending rallies by 12%. Almost three-quarters (74%) said none of these three forms of activism would have much effect.[9] The study cited "multiple polls" reporting that "Americans believe legislators ignore public opinion and act based on the wishes of campaign contributors."[10]

The result could be described as "climate despair." In one study, only 6% of respondents said society can and will reduce global warming. [11] The key barrier to climate action is not doubt about the reality and danger of climate change but doubt about the efficacy of action. The primary problem isn't denial of climate change but denial of the ability to do anything worthwhile about it. Lack of activism is largely due to a plausible (though not necessarily valid) impression of powerlessness. Climate despair must be a prime target of the climate insurgency.

Challenging Powerlessness

People who are isolated feel, and for the most part are, powerless in the face of climate change. If their only opportunities to act are alone at the supermarket and the voting booth, it is no wonder they feel there is little they can do. So the climate insurgency must draw people out of "climate isolation"—the feeling that they can only address climate change as lone individuals.

The insurgency can do this in two ways. It can provoke and inspire people to begin talking about climate change in their own milieus by confrontational actions that draw their attention and challenge them to do so. And it can reach out to people and create channels through which they can connect with those who are already active around climate change.

People's sense of powerlessness is a product both of their actual experience and of the deliberate efforts of rulers and their supporters to conceal the power that people actually have. The insurgency needs to educate the people about their power. It needs to explain the basic concept that the power of the authorities depends on the cooperation and acquiescence of the apparently powerless—and that such cooperation and acquiescence can be withdrawn. It needs to present the many historical examples—from the civil rights, labor, women's, antinuclear, and many other movements—that demonstrate the power of the people. And it needs to show the specific cases, like the defeat of the Keystone XL pipeline, that show the power of the people can directly address and defeat the causes of global warming.

Describing isolated victories is not enough. People can easily see that blocking pipelines, power plants, and fossil fuel trains one at a time will not halt climate change in the necessary timeframe. The movement needs to present a strategy that shows how people's actions can be amplified into a force that can overcome the fossil fuel forces. The strategy laid out in this book is intended as a contribution to that effort, and it can no doubt be improved by many other creative contributions.

The most effective way to overcome the belief in powerlessness is to let people observe and experience exemplary actions that reveal and prove the power of the people. Well-organized mass nonviolent direct action is a powerful way to demonstrate that power. Such actions need to show that people can act and that their action can have an impact. That impact can take the form of actually affecting power-holders, for example by forcing them to cancel a fossil fuel infrastructure project. But it can also take the form of changing the beliefs and behavior of the public. Actions will have additional impact if they can be seen as part of a trajectory which, with sufficient growth, could actually put an end to fossil fuel emissions.

Action begets a sense of empowerment, which in turn begets further action. In 2013, forty thousand people demonstrated in Washington, DC, for climate protection policies. It was the largest climate protest in American history up to that time. A year later, four hundred thousand people joined the People's Climate March through New York City. It would be difficult to prove, but also difficult to deny, that the tenfold increase in participation had something to do with the dramatic struggles over the Keystone XL pipeline that had occurred in the intervening year.

Chapter 10

Undermining the Pillars of Support for Climate Destruction

THE FOSSIL FUEL INDUSTRY COULD NOT FUNCTION FOR A day without the support and acquiescence of other forces. Police, prosecutors, judges, juries, and ultimately the military protect its "right" to use its property to destroy the climate. Corporations that are themselves threatened by climate change nonetheless use their wealth and power to promote climate destruction policies through bodies like the U.S. Chamber of Commerce and the Republican Party. Some labor unions fight climate protection policies like the Clean Power Plan and support fossil fuel infrastructure projects like the Dakota Access pipeline on the grounds that doing so will be good for jobs. Many religious groups are largely silent on climate destruction, even though it represents the most morally abominable act in human history; as a result, their members feel comfortable not taking a stand against fossil fuels.

While nonviolent movements may engage directly in dramatic confrontations, the secret of their success is often an indirect strategy that, instead of overwhelming their opponents head-on, undermines their power by removing what are often referred to as their "pillars of support." Without the support and acquiescence of law enforcement, other businesses, labor unions, religious groups, and similar institutions the power of the fossil fuel forces would crumble.

Undermining the pillars of support for fossil fuels is something for people to pursue in every institution and every milieu of which they are part. Encouraging all kinds of people and groups to do so is a strategic objective of the climate insurgency. This chapter just gives a few examples.

Constituencies

In 1988, even before preliminary climate protection negotiations began, the U.S. oil and auto industries and the U.S. National Association of Manufacturers established the Global Climate Coalition to oppose any mandatory actions to address global warming. Housed at the National Association of Manufacturers, the coalition spent tens of millions of dollars on advertising against international climate agreements and national climate legislation.[1]

Since that time, the fossil fuel industry's hegemony over the rest of business has gradually eroded as global businesses shifted away from blanket opposition to climate protection measures. In the lead-up to the Copenhagen climate summit, many large corporations withdrew from the Global Climate Coalition and joined the newly formed Business Environmental Leadership Council, which endorsed climate science and supported binding international agreements for climate protection, albeit ones with a business-friendly approach.[2]

In some cases, movement action has deliberately divided business opposition to climate protection. The U.S. Chamber of Commerce has been a significant vehicle for the fossil fuel industry to exercise hegemony over the rest of American business on climate issues. As some firms began pulling out of the Chamber because of its climate policies, 350. org launched a campaign called The U.S. Chamber Doesn't Speak for Me.[3] When the leading climate denialist Heartland

Institute posted a billboard with a picture of Unabomber Ted Kaczynski and a text reading "I still believe in Global Warming. Do you?" 350.org and other groups started a petition demanding that the ad be withdrawn, the Heartland Institute apologize, and corporations cease funding its work.[4] Many companies withdrew from both the Chamber and the institute.

The fossil fuel divestment campaign has had a significant if hard-to-measure impact on business. It has led some businesses and investors to divest—whether out of support for climate protection or to protect their investments against becoming "stranded assets."

Taking another approach to shifting business climate policy, tycoons Michael R. Bloomberg, Henry Paulson, and Tom Steyer organized the Risky Business Project, which has issued a series of national and regional reports on the threat of climate change to American business. Widely distributed and heavily promoted, these reports have received wide media attention.

While the fossil fuel industry has often exercised hegemony over the rest of the business community, that does not mean that their interests are indeed the same. Corporations are primarily interested in profit. Persuading them that climate change will be catastrophic for their future, and that eliminating fossil fuels can contribute to their profitability, can help remove them as a pillar of support for fossil fuels.

While these business forces often present inadequate solutions to climate protection, their defection from fossil fuel hegemony changes the playing field and reduces the relative power of the fossil fuel forces. Merely neutralizing them is not sufficient, but it can be an important step. Continuing that process is a strategic objective for the climate insurgency.

The AFL-CIO opposed the Kyoto climate agreement and never supported the failed Copenhagen Agreement.

However, a variety of forces have been striving to help organized labor change its approach to climate protection. The Blue-Green Alliance focuses on green jobs as a way to make climate protection of immediate interest to labor. Trade Unions for Energy Democracy encourages labor support worldwide for public ownership of utilities and other energy corporations. The Labor Network for Sustainability works to bring unions into a wider agenda for economic, social, and environmental sustainability, and to campaign for a just transition for workers threatened by the elimination of fossil fuels. In 2016, LNS initiated a Labor Convergence on Climate that is organizing around a climate agenda across the labor movement. While the AFL-CIO opposed past climate agreements, it applauded the Paris Agreement as "a landmark achievement in international cooperation" and called on America "to make the promises real."[5]

Evangelical Christianity, often allied with political conservatism, has been a bastion of climate denialism and opposition to climate protection policies. A growing sector of evangelicals has challenged this approach, however. The Evangelical Climate Initiative, formed in 2006, released a statement making a moral argument for climate action. Dozens of evangelical leaders signed, including Rick Warren, Leith Anderson, and Joel Hunter, whose megachurches have tens of thousands of members. The movement proclaims the reality of climate change and calls for federal legislation to lower greenhouse gas emissions. In its first six years the group grew from about fifteen thousand members to over eight hundred thousand, with an aim of reaching three million more.[6]

These are only a few examples of the many constituencies that have sometimes provided support or acquiescence to the fossil fuel industry, but that have forces "boring from within" that are eroding that support. Similar examples could be given for parents, students, educators, journalists, academics, scientists, and other constituencies.

Politics

Denial of climate change by conservative political forces concentrated within the right wing of the Republican Party has been a crucial pillar supporting fossil fuel industry policies. It has succeeded in blocking virtually all climate protection legislation and policy in Congress. There are forces that are undermining this pillar, however, and the climate insurgency can help encourage them.

The Republican Party is not a monolith on climate issues. As we saw in Chapter 8, half of liberal and moderate Republicans are worried about global warming.[7] One in three conservative Republicans thought reaching an agreement on global warming in Paris was important.[8] And 37% of conservative Republicans support setting strict limits on carbon dioxide emission.[9]

Further, opposition to climate protection is extremely unpopular with the American electorate. Asked in a recent poll if they would be more, less, or as willing to vote for a presidential candidate who strongly opposes action to reduce global warming, registered voters are more than three times more likely to vote against such a candidate (43%) than for them (13%).[10] How could Donald Trump have won the 2016 presidential election despite this? Perhaps it has something to do with the fact that not a single question about climate policy was asked in the three presidential debates.[11]

The climate insurgency can play a critical role in defining what constitutes "action to reduce global warming." It does not need to engage in wrangling over the details of climate policy. But it can lay out clear criteria for judging political candidates—criteria that are in line with its own basic policy.

It can insist that to be judged a climate protector, a candidate must oppose all new fossil fuel infrastructure and propose and support Climate Action Plans that will eliminate fossil fuel burning by 2050 at the latest.[12] No doubt some allies will support candidates who do not meet this test, but the insurgency's clear statement of what is necessary can establish a standard that frames the debate.

The Forces of Law and Order

The climate insurgency is already seeing signs of sympathy and support from those whose jobs involve protecting the climate destroyers. When protesters block fuel trains or occupy government buildings, normally the police are called and the protesters are arrested and tried as lawbreakers. But a trickle of recent climate cases has begun to erode the expectation that the law supports the right of property owners to use their property to destroy the climate.

A prototypical example occurred in 2007 when five activists tried to shut down a coal-fired power plant in Kent, England. The protesters admitted they had tried to shut down the plant, but argued that they were legally justified in doing so because they were trying to prevent climate change from causing far greater damage to property around the world.

In an eight-day trial, NASA's James Hansen told the court that carbon dioxide emitted from the plant could be responsible for the extinction of up to four hundred species; other witnesses testified that climate change was already causing local flooding in Kent. In his summing-up at the end of the trial, the judge said the case centered on whether or not the protesters had a "lawful excuse" for their actions. He told the jury of nine men and three women that to use a lawful excuse defense the protesters had to prove that their action was due to an immediate need to protect property belonging to another. By majority vote the jury voted to clear the Kingsnorth Six.[13]

No similar case has yet occurred in the U.S., but the legal system is producing more and more expressions of sympathy for climate protesters. For example, on Earth Day 2013, Alec Johnson (a.k.a. "Climate Hawk") locked himself to a construction excavator in Tushka, Oklahoma, as part of the Tar Sands Blockade campaign to stop the Keystone XL pipeline. In a statement he prepared for the jury, Alec Johnson argued that his blockade was necessary because the pipeline threatens our atmospheric public trust and state and national governments are failing to protect us against that threat.[14] He proclaimed on the basis of the public trust principle, "I wasn't breaking the law that day—I was enforcing it."[15] Although Johnson could have been sentenced to up to two years in the Atoka County Jail, the jury ordered no jail time and a fine of just over $1,000.

In 2013, Jay O'Hara and Ken Ward used a small fishing boat named *Henry David T* to block a ship from unloading forty thousand tons of coal at the Brayton Point, Massachusetts power plant. Prosecutors charged them with disturbing the peace, conspiracy, failure to act to avoid a collision, and negligent operation of a motor vehicle. O'Hara and Ward argued that the imminent threat of global climate change left them no choice but to act as they did. The day the trial was set to begin, the Bristol County district attorney went out to the steps of the courthouse and announced that he was reducing the charge to a modest fine, which would help defray municipal costs. Then he issued a statement in support of O'Hara and Ward's protest. "Climate change is one of the gravest crises our planet has ever faced. In my humble opinion the political leadership on this issue has been gravely lacking." He thereupon met with the defendants and told them he would join them at the upcoming People's Climate March.

On September 2, 2014, five activists blockaded a train used to ship Bakken oil in a BNSF Delta rail yard in Everett, Washington. The "Delta 5" argued that "to seriously address the climate crisis, we need to be shutting down our fossil fuel infrastructure and keeping that oil in the ground." On that basis they maintained that their blockade was "morally—and legally—justifiable given the imperatives of the climate crisis." They asked that their actions be viewed, "not as a crime, but as a reasonable act of conscience, necessitated by the extreme nature of the emergency and by the fact that the government itself is in violation of the law."

Initially the judge refused to admit a necessity defense. But shortly before the trial he reversed himself. As a result, for the first time in U.S. history a judge allowed a jury to hear testimony that climate protesters should not be found guilty of breaking the law because their actions were necessary to prevent a far greater harm—destruction of the earth's climate. After testimony was presented, however, the judge instructed the jury not to consider the necessity defense. But the jury had already heard why the Delta 5 did what they did—and the expert testimony on the threat presented by climate change and oil trains. The jury acquitted the Delta 5 on the major charge of obstructing a train and found them guilty only of trespass. At the end of the trial three of the jurors met with the defendants in the hallway, hugged them, and agreed to join them for an upcoming climate lobby day. They said that but for the judge's firm instructions they would have voted to acquit.

The day after the 2014 People's Climate March, several thousand "Flood Wall Street" protesters blocked traffic in New York's Financial District for much of the day. Approximately one hundred people were arrested, of whom ten faced trial and were found not guilty. New York City Criminal Court Judge Robert Mandelbaum ruled that the dispersal order issued by the New York Police Department constituted an unlawful violation of demonstrators' First Amendment rights. Judge Mandelbaum also took judicial notice of the fact that climate change is real, human-made, and requires drastic action. Defense attorney Martin Stolar said that this acknowledgment is "unprecedented and has significance for future litigation involving climate change."[16]

Albany Break Free from Fossil Fuels organizers had met with officials from the police and sheriff's offices and reported, "They abhor the trains and are very supportive of us." These authorities told organizers there would be no mass arrests, and despite a day-long occupation of railroad property, they kept their word.

The legal system is still capable of meting out draconian punishments to climate protesters. Tim DeChristopher, for example, was forced to serve twenty-one months in federal prison for the alleged "crime" of bidding at a gas lease auction without intending to pay. But at the Break Free action in Lakewood, Colorado, hundreds of people disrupted an auction of oil and gas drilling rights; seven held a sit-in blockading the room where the auction was being held; but not a single protester was arrested.

Nobody should commit civil disobedience in the expectation that they will be acquitted on constitutional or public trust grounds. But these cases show that we can expect a growing proportion of our neighbors—including some who serve as police officers, sheriffs, judges, and jury members—to recognize that climate change must be halted by all means necessary and that our actions are justified to hasten that result.

In all of these cases, the "weakening of the pillars" does not equate to full support for a fossil freeze and commitment to a plan to eliminate fossil fuels by or before 2050. It is the role of the insurgency to continue making clear that such changes are necessary. However, the crumbling of the pillars of support for fossil fuel is a critical contribution to realizing them. And it is a reminder that our efforts, far from being futile, have already had significant effect. It is the slow, barely visible erosion of the pillars of support that in the end can cause them—and the fossil fuel forces they support—to fall.

Chapter 11
Just Transitions

THE ELIMINATION OF FOSSIL FUELS AND THEIR REPLACE-ment by clean energy offers potential benefits but also potential threats. As part of its effort to win support of the public, the climate insurgency needs to propose and implement ways both to realize the benefits and to forestall the threats.

The transition from fossil fuels requires a planned, rapid construction of renewable energy infrastructure on a massive scale, comparable to but greater than the economic mobilization that the United States undertook during World War II. By itself the market will not provide such a transformation; it will require public planning and investment designed to protect the climate for the public good.[1]

A fossil freeze threatens the jobs of those workers who produce and use fossil fuel and the construction workers who build fossil fuel infrastructure. A fossil freeze campaign will need a jobs program that includes both a just transition for workers who will be affected by the freeze and a way to take advantage of the transition to a fossil free economy to create millions of jobs for all kinds of workers, especially those who have previously been the victims of discrimination and exclusion from good jobs.

Fortunately, the primary way to reduce GHGs is to replace fossil fuels with renewable energy and energy efficiency—and that produces far more jobs than fossil fuel energy.

In 2015, the Labor Network for Sustainability (LNS) and 350.org issued a report, "The Clean Energy Future: Protecting the Climate, Creating Jobs, and Saving Money."[2] It shows that the U.S. can reduce GHG emissions 80% by 2050—while adding half-a-million jobs and actually cutting utility bills compared to business-as-usual fossil fuel energy. Most of the added jobs will be in manufacturing and construction.

The plan does not depend on any new technical break-throughs to realize these gains, only a continuation of current trends in energy efficiency and renewable energy costs. It is based on the conversion of all gasoline powered light vehicles and most space heating and water heating to 100% renew-able electricity. It includes an orderly phasing out of coal and nuclear energy and a gradual reduction in the burning of natural gas.

In addition to such national plans, campaigns against fossil fuel infrastructure need to propose specific alternatives for the pipelines, plants, and other facilities they are trying to block or close. These need to address both consumers' need for energy and workers' need for jobs. An example: when a 2015 campaign initiated by Baltimore high school students blocked a large incinerator project, the students organized a celebratory Concert for Fair Development, highlighting proposals for a solar farm on the site, zero waste reuse and recycling industries, and local agricultural initiatives.[3]

In the Keystone XL battle, "jobs vs. the environment" was a central theme. In response, the Labor Network for Sustainability issued the report "The Keystone Pipeline Debate: An Alternative Job Creation Strategy."[4] The report shows how to create five times as many jobs as the Keystone XL pipeline by investing in much-needed water, sewer, and gas infrastructure maintenance and repair in the five states along the proposed pipeline route. The study found that meeting water and gas infrastructure needs in those five states alone would create the equivalent of three hundred thousand jobs, each lasting a year. Every dollar spent on gas, water, and sewer infrastructure in those states would generate 156% more employment than the proposed Keystone XL pipeline.

The "Clean Energy Future" report found that the transition to clean energy will produce five times as many jobs as it eliminates. That is cold comfort, however, for anyone whose job is threatened by that transition. Climate protectors need to ensure that "no worker is left behind" by the transition to a climate-safe economy.

Break Free from Fossil Fuels emphasized the need for such a just transition. Break Free–Midwest, for example, drew up and passed a formal "Why We March" resolution, which stated:

> Just Transition includes subsidized support of the workers who have labored in the fossil fuel industry, providing them with meaningful counseling and financial support for the workers and their families as they acquire the training and development of new skill sets to work in the new Green Economic Industrial Corridor (GEIC), a Just Transition for workers that will include new jobs in the renewable energy industry, the creation of new conservation efforts to sustainably retrofit residences and commercial properties, repairing our decaying infrastructure, and create new, efficient systems of public transportation.[5]

An example of how current fossil fuel workers might be protected is the "Clean Energy Worker Just Transition Act" proposed by Sen. Bernie Sanders (I-Vt.), Jeff Merkley (D-Ore.), and Edward Markey (D-Mass.).[6] The bill initially targets coal workers, but over time expands to other energy sector workers as well. It provides unemployment insurance,

health care, and pensions for up to three years and job training and living expenses up to four years. Employers receive
tax incentives to hire transitioning employees. Counties
where thirty-five or more workers become eligible for the
program can receive targeted development funds. The right
of workers to join unions is protected by streamlining NLRB
union recognition provisions. The bill covers the estimated
$41 billion cost of the program by closing the tax loophole
that allows corporations to send their headquarters overseas
to avoid paying taxes.

Good, stable jobs protecting the climate can help challenge the growing inequality and injustice of our society, but
only if public policy is designed to do so. Climate policy needs
to include strong racial, gender, age, and locational hiring
requirements to counter our current employment inequality
and provide a jobs pipeline for those individuals and groups
who have been denied equal access to good jobs. It needs to
help remedy the concentration of pollution in low-income
communities; their lack of transportation, education, health,
and other facilities; and other manifestations of discrimination, inequality, and racism.

The climate insurgency can be pluralistic about what
kind of future society should follow the elimination of fossil fuels. But it can open the way to wide discussion of and
experimentation with new social forms. It can help draw together a community of allies who support climate protection,
even though it is not their primary focus. It can help them
develop a common program through which their needs and
objectives can be met as part of the transition to a climate-
safe society.[7]

Chapter 12

The Right of the People to Protect the Climate

THOUSANDS OF PEOPLE HAVE ENGAGED IN CIVIL DISOBEDIence for climate actions like Break Free from Fossil Fuels. Many have been arrested and faced legal consequences.

Many of them have begun to define themselves—to the movement, the public, and the courts—not as criminals but as law-enforcers trying to protect legal rights and halt governments and corporations from committing the greatest crime in human history. As the Break Free Proclamation put it, "We proclaim: The people of the world have a right, indeed a duty, to protect the public trust we own in common—the earth's climate. When we take nonviolent direct action we are law-enforcers carrying out our duty to protect the earth's climate from illegal, dangerous crimes." That assertion defines the climate insurgency.[1]

What Is the "Public Trust Doctrine"?

The governments that permit and protect climate destruction may rule the world, but they do not own the world. Under many legal systems the earth's shared natural resources belong to the world's people and their posterity, as the common heritage of humanity.

Governments have long served as trustees for rights held in common by the people—specifically, rights to the public natural resources on which we all depend. In American law this role is defined by the "public trust doctrine," under which our federal and state governments serve as trustees of natural resources on behalf of present and future generations. As trustee the government has a strict "fiduciary duty" to the owners—the citizen beneficiaries. This legal duty requires government officials to act in the public interest, with "the highest duty of care." Our officials have no legal right to harm the public trust for the benefit a corporation or other private interest—no matter how politically powerful it may be.

This fundamental principle is embodied in the laws and constitutions of countries around the world. It was codified in the Institutes of Justinian, issued by the Roman emperor in 535 AD, which stated, "By the law of nature these things are common to mankind—the air, running water, the sea, and consequently the shores of the sea."

Use of the public trust doctrine for climate protection has been pioneered by young people represented by Our Children's Trust, who have brought lawsuits and rulemaking petitions in every U.S. state, against the federal government, and in countries around the world, to require governments to act on their public trust duty to protect the climate and to protect fundamental constitutional rights.

Xiuhtezcatl Martinez, one of the lead youth plaintiffs in the federal case, and youth director of Earth Guardians, another plaintiff in the federal case said:

> The Federal government has been making decisions
> in the best interest of multinational corporations
> and their profits, but not in the best interest of my
> generation and those to come. Instead of changing
> their business model to meet the scientific reality of
> climate change, these companies are demanding we

adapt to an uninhabitable world that supports their profits. When you compare the two, I think it's clear that our right to clean air and a healthy atmosphere, is more important than their "need" to make money off destroying our future.

In November 2015, attorneys for American Fuel & Petrochemical Manufacturers, American Petroleum Institute, and the National Association of Manufacturers, filed a motion to "intervene" and join forces with the government as codefendants in the federal constitutional and public trust lawsuit brought by Our Children's Trust. They argued, "If Plaintiffs succeed in this Court ordering the elimination or massive reduction of U.S. conventional fuel consumption and manufacturing processes that emit GHGs beyond existing federal and other regulations, the members of each of the Proposed Intervenor-Defendants will be harmed." Two months later Magistrate Judge Thomas Coffin of the U.S District Court in Oregon accepted the fossil fuel and manufacturing industries' move to join the federal government in opposing the climate kids' lawsuit.

"The fossil fuel industry would not want to be in court unless it understood the significance of our case," said Philip Gregory, attorney for the youth. "This litigation is a momentous threat to fossil fuel companies. They are determined to join the federal government to defeat the constitutional claims asserted by these youth Plaintiffs. The fossil fuel industry and the federal government lining up against 21 young citizens. That shows you what is at stake here."

The right of the people to demand that government protect the public trust expresses a common-sense constitutionalism. In the United States, we enjoy fundamental constitutional rights to life and liberty. Government failure to protect the public trust, and governments' complicity in the destruction of the atmospheric resource and the natural conditions on which human life depends, compromises those constitutional rights in the near term and extinguishes those rights in the longer term.

The Right to a Stable Climate

Two days after the election of Donald Trump, twenty-one kids won a court ruling that may be just as important as that election in determining our future. The decision by Judge Ann Aiken in the federal district court in Oregon sets the stage for a momentous trial of the right to a stable climate—and the constitutional obligation of the United States government to protect that right. It also lays out the arguments that justify climate insurgency.

As Judge Aiken emphasized, "This is no ordinary lawsuit." The youth's suit, supported by the grassroots, youth-focused climate recovery nonprofit Our Children's Trust,[2] challenges decisions "across a vast set of topics"—decisions like "whether and to what extent to regulate CO_2 emissions from power plants and vehicles, whether to permit fossil fuel extraction and development to take place on federal lands, how much to charge for use of those lands, whether to give tax breaks to the fossil fuel industry, whether to subsidize or directly fund that industry, whether to fund the construction of fossil fuel infrastructure such as natural gas pipelines at home and abroad, whether to authorize new marine coal terminal projects."

Judge Aiken noted the personal harms the kids said they face because of climate change. One said the algae blooms harm the water she drinks, and low water levels caused by drought kill the wild salmon she eats. Another said increased wildfires and extreme flooding jeopardize his personal safety. Jayden F., a thirteen-year-old resident of Rayne, Louisiana, said that at five o'clock on the morning of August 13, 2016, she stepped out of bed into ankle-deep water.

> Floodwaters were pouring into our home through every possible opening. We tried to stop it with towels, blankets, and boards. The water was flowing down the hallway, into my Mom's room and my sisters' room. The water drenched my living room and began to cover our kitchen floor. Our toilets, sinks, and bathtubs began to overflow with awful smelling sewage because our town's sewer system also flooded. Soon the sewage was everywhere. We had a stream of sewage and water running through our house.

Jayden said the storm that destroyed her home "ordinarily would happen once every thousand years, but is happening now as a result of climate change."

The Fifth Amendment to the U.S. Constitution bars the federal government from depriving a person of "life, liberty, or property" without "due process of law." The climate kids say that the effects of climate change they describe do just that—and that the policies of the federal government are responsible for those violations of their rights.

The lawsuit alleges that the government has violated their rights by "directly causing atmospheric CO_2 to rise to levels that dangerously interfere with a stable climate system." The government knowingly endangered their "health and welfare" by "approving and promoting fossil fuel development," including "exploration, extraction, production, transportation, importation, exportation, and combustion." And after "knowingly creating this dangerous situation," it continued to "knowingly enhance that danger" by "allowing fossil fuel production, consumption, and combustion at dangerous levels." These government decisions have caused the planet to warm and the oceans to rise. There is a direct causal line between the government's policy choices and floods, food shortages, destruction of property, species extinction, and a host of other harms.

The climate kids said these policies not only violate their individual constitutional rights but also transgress the duty of the government to preserve the core natural resources necessary for the well-being and survival of the people. Those resources constitute the "public trust." They are legally protected as the common property of the people. The climate kids' suit says that the government has violated its duty as trustee of the public trust by allowing the depletion and destruction of the atmosphere—thereby violating the common property rights of the public and future generations.

The government acknowledged in court that climate change poses "a monumental threat to Americans' health and welfare" by "driving long-lasting changes in our climate," leading to an array of "severe negative effects, which will worsen over time." It then used a barrage of legal technicalities and irrelevant precedents to claim that the climate kids don't have legal standing to bring such a suit; that climate change is a "political question" that courts should leave to other branches of government; and that anyway courts don't have the power to halt climate change.

Judge Aiken's decision cut through this smokescreen to focus on the essential point: "The right to a climate system capable of sustaining human life is fundamental to a free and ordered society." A stable climate system is quite literally the foundation of society, "without which there would be neither civilization nor progress."

The youth frame the fundamental right at issue as "the right to a climate system capable of sustaining human life." If "governmental action is affirmatively and substantially damaging the climate system in a way that will cause human deaths, shorten human life spans, result in widespread damage to property, threaten human food sources, and dramatically alter the planet's ecosystem," then the youth have a claim for protection of their life and liberty under the Fifth Amendment. "To hold otherwise would be to say that the Constitution affords no protection against a government's knowing decision to poison the air its citizens breathe or the water its citizens drink."

Judge Aiken also allowed the climate kids' public trust case to go forward, quoting a judicial opinion that the right of future generations to a "balanced and healthful ecology" is so basic that it "need not even be written in the Constitution" for it is "assumed to exist from the inception of humankind."

The climate kids asked the court to declare that their constitutional and public trust rights have been violated and to order the government to develop a National Climate Recovery Plan to reduce emissions to a climate-safe level. The case will no doubt be fought out for years all the way to the Supreme Court. But the advocates of climate protection need not wait to champion the basic principles enshrined in Judge Aiken's decision. We can start doing so right now—by climate insurgency.

Using the Public Trust Frame

Claims that government actions are illegal and unconstitutional have played an important role in empowering social movements. They strengthen participants by lending a sense of clarity that they are performing a public duty. And they strengthen a movement's appeal to the broader public by presenting action not as wanton lawbreaking, but as an effort to rectify governments and institutions that are themselves in violation of the law.

For the civil rights movement, the U.S. Constitution's guarantee of equal rights meant that those engaged in sit-ins and freedom rides were not criminals but rather upholders of constitutional law—even if southern sheriffs threw them in jail. For the activists of Solidarity, the nonviolent revolution that overthrew Communism in Poland was not criminal sedition but an effort to implement the international human and labor rights laws ratified by their own government.

The constitutional duty of governments to protect the public trust, and the right of the people to life and liberty, can play much the same role in the climate movement that the U.S. Constitution's right to equality played for the civil rights movement and the Polish government's legal commitment to human and labor rights played for Solidarity.

Constitutional and public trust principles make it possible for the climate insurgency to turn the tables on the governments that purport to represent the world's people and to have the authority to rule the world. They stand for the proposition that governments do not have the right to destroy the climate—and that the people have the right to stop them when they do so.

Governments have no more right to authorize the emission of greenhouse gases that destroy the climate than the trust officers of a bank have to loot the assets placed under their care. The people of the world have a right to our common natural resources. And we have a right, if necessary, to protect our common assets against those who would destroy them.

Those who perpetrate climate change, and those who allow them to do so, should not be able to claim that the law is on their side. Those who blockade coal-fired power plants or sit down at the White House to protest fossil fuel pipelines can—and should—insist that they are exercising their fundamental constitutional rights to life and liberty and their responsibility to protect the atmospheric commons they own along with all of present and future humankind. Climate protesters can proudly proclaim that they are actually upholding the law, protecting constitutional and public trust rights for all.

The idea of using the public trust doctrine to legitimate climate action is spreading. In February 2016, 350.org's organizers for the Break Free from Fossil Fuels actions in the U.S. agreed to use the public trust doctrine to frame U.S. Break Free actions. A working group was established to prepare materials and resources to bring the public trust frame into actions around the U.S. As we have seen in Chapter 1, the public trust became a central part of the Break Free narrative in Albany and elsewhere.

The climate insurgency asserts that nobody has the right to destroy the climate system on which the present and future of humanity and life itself depend. Governments have no higher duty than to protect the earth's climate on behalf of the earth's people. If they fail to do so, the people have a right to protect it themselves. The climate insurgency denies the sovereignty, legitimacy, and authority of those governments that are violating the rights to life, liberty, and the public trust by authorizing and encouraging the destruction of the earth's climate system. The public trust frame can "flip the script" in civil disobedience and nonviolent direct action, making it clear that the fossil fuel industry and the governments that do its bidding are criminals, while the climate insurgency is upholding the Constitution and the public trust.

The people's withdrawal of legitimacy is the ultimate threat, the ultimate sanction, the "nuclear option." It creates a situation in which millions of people won't stay off the fossil fuel companies' private property, and in which police won't arrest, prosecutors won't prosecute, and judges and juries won't convict—and the public will support the protesters' defiance of purported law. Hopefully the threat of "inconvenience" will take effect long before the nuclear option. But if those who claim the right to destroy humanity's future decline to halt their destruction, they should know what to expect.

Chapter 13
Dual Power

As world leaders descended on the United Nations in the aftermath of the 2014 People's Climate March, across the street representatives of peoples impacted by climate change from around the globe assembled for a People's Climate Justice Tribunal. On the basis of their testimony a people's judicial panel found governments and corporations culpable of climate destruction and defended those who try to halt it. Such tribunals can become a critical part of the climate insurgency, passing judgment on the responsibilities of polluters, authorizing action to protect the climate, and serving as an alternative center of legitimate authority.[1]

After hearing the testimony from people around the world, the tribunal's judicial panel of respected movement figures, citing the public trust doctrine, addressed the responsibility—and the culpability—of governments:

> The governments of the world have a duty to protect the atmosphere that belongs in common to the world's people. Based on the evidence we have heard here today, the nations of our world are in violation of their most fundamental legal and constitutional obligations. They are violating the most fundamental rights of their own people and the people of the world. Each government should be legally compelled to halt its contribution to climate destruction.

The panel recognized that it does not have the authority to force governments to take such action. But it cited the view of professor of international law Richard Falk, who observed in the opening speech at the World Tribunal on Iraq, "When governments and the UN are silent and fail to protect victims of aggression, tribunals of concerned citizens possess a law-making authority." The judges argued the same is true when governments fail to protect victims of climate change.

The judicial panel observed that failure of governments to protect human rights and the public trust has prompted the co-owners of the atmospheric commons to turn to mass civil disobedience to protect their common property. These actions must be seen "not as violations of the law, but as attempts to enforce it." They represent the effort of tens of thousands of people to assert their collective right and responsibility to protect our public trust property, which includes resources like the earth's climate that are essential to our survival.

> Based on the evidence we have heard today, those who blockade coal-fired power plants or block tar sands oil pipelines are committing no crime. Rather, they are exercising their right and responsibility to protect the atmospheric commons they own along with all of present and future humankind. They are acting to prevent a far greater harm—indeed, a harm that by virtue of the public trust doctrine is itself a violation of law on a historic scale.

Climate Tribunals and the Power of the People

In 1967, Bertrand Russell and Jean-Paul Sartre, undoubtedly the most famous philosophers of their era, convened the International War Crimes Tribunal, in which a distinguished panel heard evidence that the United States was committing war crimes in Vietnam. The tribunal led millions of people around the world to question the Vietnam War and encouraged tens of thousands to resist it. It has inspired many civil society tribunals since, including more than twenty independent international tribunals held in countries around the world to examine the criminality of the Iraq war. The People's Climate Justice Tribunal was neither as prestigious nor as ambitious as the Russell Tribunal, but it continued that tradition.

The People's Climate Justice Tribunal established a credible case that the governments and corporations of the world are systematically violating human rights, international law, and their duty to protect the public trust by allowing the greenhouse gas emissions that are destroying the earth's climate. Future climate tribunals could examine the evidence in greater detail. They could issue declaratory judgments and injunctions. They could also make findings on the rights and responsibilities of global citizens to enforce the law and their legal rights vis-à-vis governments that try to subdue them when they do so.

Tribunals can be convened as part of the legitimation and public education activities of specific campaigns. Although they may be initiated by the insurgency, the validity of their judgments can be based on the fairness of their conclusions and the evidence and argument on which they are based. Some tribunals could become permanent institutions.

In specific instances, people could apply to such tribunals for "advisory opinions" on questions like the need to halt new fossil fuel infrastructure or the adequacy of Climate Action Plans. Tribunals could weigh the evidence and issue judgments. They could then negotiate consent decrees or issue advisory orders. The people could then attempt to impose or implement those orders by mass action.

Such orders could draw on the legal principles that are emerging from the climate kids' lawsuits. For example, tribunals could take as a model the Washington court decision that ordered the state to implement a Climate Action Plan on the basis of a lawsuit by Our Children's Trust.[2] And it could order the kinds of remedies proposed in the climate kids' federal lawsuit:

> This Court should order Defendants to cease their permitting, authorizing, and subsidizing of fossil fuels and, instead, move to swiftly phase out CO_2 emissions, as well as take such other action as necessary to ensure that atmospheric CO_2 is no more concentrated than 350 ppm by 2100, including to develop a national plan to restore Earth's energy balance, and implement that national plan so as to stabilize the climate system.[3]

Such tribunals might influence official courts to enforce the law against fossil fuel corporations and governments that are acting like their captives. And if they do not, people's tribunals may have a role to play in legitimating nonviolent insurgency on the part of the people of the world to save planet earth.

Governmental Rebellion

Climate insurgency can even spread to parts of the government itself.[4] In Deerfield, Massachusetts, in 2016, the Texas-based Kinder Morgan company asked the Massachusetts Department of Public Utilities to force the more than four hundred property owners along the route of its proposed Kinder Morgan natural gas pipeline to allow company surveyors on their land. In reply, the town of Deerfield wrote the DPU that its Board of Health had forbidden all activities of Kinder Morgan in the town. The health board had said that "a corporation convicted of felonies resulting in the tragic deaths of five people presents an unreasonable risk to the health and lives of residents of Deerfield if such felon were to be allowed to build a massive fracked gas pipeline through the town."

The Select Board of the town warned that anyone entering onto private properties, without permission from the property owners, for activities related to the proposed natural gas pipeline will be arrested for trespassing—even if they have an order from the DUP. A lawyer representing the town said Deerfield is "prepared to supersede any state authority and have police officers arrest anyone who enters onto private property as part of the pipeline project." Kinder Morgan claimed the federal Pipeline Safety Act preempts any state's authority to regulate pipeline safety and that certain state laws trump the town's orders. In the face of massive opposition, a few months later Kinder Morgan withdrew its plan for the $3 billion pipeline.

In 2013, the Pennsylvania General Energy Company (PGE) applied for permits for wells to inject contaminated fracking wastewater in Grant Township, Pennsylvania. Despite hearings, public comments, and permit appeals, the federal Environmental Protection Agency issued a permit to PGE. With support from the Community Environmental Legal Defense Fund, the Grant Township Supervisors thereupon passed a Community Bill of Rights ordinance which established rights to clean air and water, community self-government, and the rights of nature; it prohibited the proposed injection well as a violation of those rights.

PGE sued Grant Township, claiming it had a right to inject wastewater within the Township. When a judge stated that the Township did not have the authority to prohibit injection wells, the residents voted two to one for a new home rule charter. Then the Grant Township Supervisors passed a law under the new charter that legalizes direct action to stop fracking wastewater injection wells. It states that if a court does not uphold the people's right to stop corporate activities threatening the well-being of the community, "any natural person may then enforce the rights and prohibitions of the charter through direct action." Further, the ordinance prohibits "any private or public actor from bringing criminal charges or filing any civil or other criminal action against those participating in nonviolent direct action."

Grant Township Supervisor Stacy Long said, "I live here, and I was also elected to protect the health and safety of this Township. I will do whatever it takes to provide our residents with the tools and protections they need to nonviolently resist aggressions like those being proposed by PGE." Grant Township Supervisor and Chairman Jon Perry added, "I was elected to serve this community, and to protect the rights in our Charter voted in by the people I represent. If we have to physically and nonviolently stop the trucks from coming in because the courts fail us, we will do so."[5]

Years before the American Revolution, the British government imposed a new set of taxes on their American colonies, and the colonists responded with a boycott of British goods, codified in a "Nonimportation Agreement" and a Nonimportation Association to back it. Association committees held hearings, took testimony and examined the records of those suspected of violating the agreement, judged their guilt, and imposed sanctions on violators—much like courts of law. Those found guilty were subjected to social ostracism and visits by angry crowds. Public opinion seemed to treat the Nonimportation Agreement as more legitimate than the official government; one royal governor complained that tea smuggled from Holland could "lawfully be sold" in Boston, whereas it was considered "a high crime to sell any from England."[6]

The threat of such a "dual power" may provide the climate insurgency's ultimate sanction.

Conclusion
Two Scenarios

WHAT IF THE CLIMATE INSURGENCY FAILS? WHAT IF IT succeeds? Here are two scenarios.

Doom

Following the "business-as-usual" pathway, GHG emissions continue to rise, reaching 450 ppm CO_2 equivalent in the atmosphere by 2050 and 800 ppm by 2100.[1] Global temperatures rise 4.5°C (8.1°F) by 2100 over preindustrial levels—assuming no unanticipated tipping points are crossed.[2]

The probable consequences will be very much like those we have seen from climate change already, except on a greatly expanded scale. Sea level rise of up to five feet will inundate most coastal settlements and displace millions of people. Heat waves, desertification, floods, famines, water shortages, and extinctions will follow.

Following the predictions of the IPCC (echoed by U.S. military officials), climate change will cause economic decline, state collapse, civil strife, mass migrations, and resource wars.[3] Military spending will be ramped up to deal with the consequences. By 2050, climate refugees will number 250 million.[4] Social disorder and breakdowns of security will become rampant. The costs of destruction will lead to local, national, and global economic decline; according to the *Stern Review*, they will exceed all the damage of the Great Depression and World War II combined.[5]

Efforts to combat climate change will grow as populations demand protection. Despite international conferences, policy pronouncements, and much talk, coal, oil, and gas will continue to be the principal source of energy. Ever more extreme locations and technologies will be used for extraction. The places most protected from climate threats will be occupied by the wealthiest, increasing climate injustice, but even those locations will prove to provide inadequate refuge.

Advocates of climate protection will be increasingly labeled as "climate terrorists." Those identified as leaders will be jailed or assassinated; demonstrations will be violently suppressed. Armies will be mobilized to terrorize both protestors and displaced masses. Authoritarian governments will proliferate as populations rampage against unbearable conditions. Protection gangs will proliferate and in many places will become the de facto political authority.

Life will be nasty, brutish, and short. Whatever we and our posterity care about will be threatened or destroyed.

All we have to do to realize this doomsday scenario is nothing.

Fossil Free

The fossil free scenario begins exactly the same way as the doomsday scenario. Greenhouse gas emissions will continue, GHGs in the atmosphere will continue to increase, and global temperatures will continue to rise as a result of GHGs already put in the atmosphere. Floods, hurricanes, wildfires, heat waves, and other climate effects will grow more severe than today. The response to climate change, however, will be very different from the doomsday scenario.

The trickle of climate insurgency actions that began with the Keystone XL and Break Free from Fossil Fuels campaigns will become a flood of thousands of actions at fossil fuel companies, infrastructure sites, and government and corporate offices. Some will be occupied or shut down "for the duration."

Climate insurgency actions will make clear to the public and the media that burning fossil fuel causes the burgeoning extreme weather events and other climate effects and that halting it is the only solution. These dramatic actions, combined with the visible effects of climate change and wide public education, will break down the walls of silence and stimulate wide discussion in the most diverse milieus. Discussing the threat of climate change and what to do about it will become as normal as discussing the weather.

The climate insurgency and allies will block more and more proposed new fossil fuel infrastructure projects permanently. The victories of the insurgency will begin to shift the public's sense of powerlessness. Growing numbers of people will come to believe that if enough people make enough effort, the worst effects of global warming can be averted.

The insurgency's demands for adequate Climate Action Plans will be joined by action within universities, religious congregations, municipalities, states, and many other institutions and jurisdictions. Students, for example, will close their universities until Climate Action Plans that will eliminate use of fossil fuels are adopted; they will close them again if the plans aren't being implemented. Those institutions and governments that don't have Climate Action Plans will be forced to adopt them. Those that have inadequate ones will be forced to strengthen and start implementing them. Divest-invest campaigns will create massive investment pools to support the transition to clean energy.

Climate insurgency action will increasingly be accompanied by people's climate tribunals. They will establish the responsibility of governments, businesses, and institutions to fulfill their constitutional obligations to protect the public trust. The tribunals will serve initially as vehicles for public education and legitimation of movement action. Over time, however, those engaging in actions will ask the tribunals to issue judgments to halt new fossil fuel infrastructure and require adequate Climate Action Plans. Based on the evidence, tribunals will begin issuing findings, ordering infrastructure projects halted, and requiring that adequate Climate Action Plans be established. Increasingly their orders will be enforced by mass direct action. Over time, official courts too will curtail new fossil fuel infrastructure and mandate Climate Action Plans.

Police, prosecutors, judges, and juries, backed by broad public opinion, will be less and less willing to arrest, prosecute, and convict climate insurgents. Their actions will increasingly be recognized as necessary to protect the public good. This will lead to far wider participation in the climate insurgency.

Under continuous pressure from the climate insurgency and from their own constituents, forces that previously supported fossil fuel interests will begin to break away from them. Business, labor, and other pillars of support for fossil fuel use, increasingly recognizing the need to break free of fossil fuels, will institute and support Climate Action Plans that will actually do so. Public officials, candidates, and political parties will take increasingly clear stands against fossil fuel infrastructure and for adequate Climate Action Plans as the insurgency stigmatizes those who fail to do so.

Meanwhile, the global insurgency will grow to become a global superpower in its own right. Its global coordination will grow, gaining the ability to coordinate long-running transnational campaigns. It will be able to mobilize global pressure and nonviolent sanctions against every government, corporation, and institution that continues to develop new fossil fuel infrastructure or fails to develop and implement an adequate Climate Action Plan.

The climate insurgency will pressure governments to fulfill and expand their GHG reduction pledges under the Paris Agreement and to make them legally binding. The insurgency and "coalition of the willing" governments will impose nonviolent sanctions on those that refuse to do so.

All of these developments will be met with fierce push-back from the fossil fuel industry and their remaining supporters. When they discover they can no longer sell their entire program, they will create "fig leaf" programs that purport to protect the climate while allowing its destruction—and the profits to be made from that destruction—to continue far into the future. These efforts, however, will be exposed and challenged by a "political ju-jitsu" that defines such actions as themselves an outrageous abuse to the right to life and liberty of current and future generations.

As governments come to recognize that climate protection is necessary both for the future of humanity and to protect themselves against the threat of the global climate insurgency, they will begin taking measures that have previously been unthinkable. They will establish climate protection authorities like those that organized production during World War II with the power to halt fossil fuel projects, plan the conversion to zero fossil fuels, force recalcitrant corporations and institutions to obey those plans, and cut through resistance and red tape. They will engage in public investment and economic planning to accelerate climate protection by ensuring employment for everyone willing and able to work. They will require protection for those whose jobs may be threatened in the transition and opportunities for those who have been excluded from good jobs and economic advancement in the past.

Recognizing the global character of global warming, governments will establish a global fund designed to use the world's underused human and material resources to accelerate climate protection worldwide. In particular it will fund the development of fossil free energy systems and GHG drawdown in poor countries.

These efforts will put us on the track to reduce global GHG emissions to near zero by 2050. That will be enough to hold global warming below 2°C. As the benefits of doing so become apparent, the timetable for reaching net zero emissions worldwide will be further accelerated.

Then will begin the long work of restoration. Forests, farms, and human settlements will be refashioned so as to withdraw carbon and other GHGs from the atmosphere. Ocean farming will begin withdrawing nitrogen and other acids from the water. While fossil fuels will have left ineradicable scars, humanity will go as far as possible to heal the damage.

In the process of protecting and restoring the climate, something more will have been accomplished as well. Our actions will have opened the way for other forms of global cooperation and of organizing our species to provide a good life for all within sustainable limits. We will still have to mourn that which we have destroyed, but we can also take comfort in the possibilities we will open up to people of the future.

Notes

Prologue

1. Kelsey Cascadia Rose Juliana, et al. v. United States of America, et al. https://static1.squarespace.com/static/571d109b04426270152febe0/t/5824e85e6a49638292ddd1c9/1478813795912/Order+MTD.Aiken.pdf.

Introduction: Climate Insurgency vs. Doom

1. Coral Davenport, "Nations Approve Landmark Climate Accord in Paris," *New York Times*, December 12, 2015, http://www.nytimes.com/2015/12/13/world/europe/climate-change-accord-paris.html?_r=0.
2. Bruce Stokes, Richard Wike, and Jill Carle, "Global Concern about Climate Change, Broad Support for Limiting Emissions," *Pew Research Center Global Attitudes and Trends*, November 5, 2015, 4, based on 45,435 face-to-face and telephone interviews in forty countries.
3. Ibid., 5.
4. Anthony Leiserowitz et al., *Americans' Action to Limit Global Warming*, Yale Program on Climate Change Communication, http://climatecommunication.yale.edu/wp-content/uploads/2014/02/Behavior-November-2013.pdf, 37.

Chapter 1: This Is What Insurgency Looks Like

1. Oliver Milman, "'Break Free' fossil fuel protests deemed 'largest ever' global disobedience," *Guardian*, May 16, 2016, http://www.theguardian.com/environment/2016/may/16/break-free-protest-fossil-fuel; other details from https://breakfree2016.org/#locations; see also https://breakfree2016.org/press-release/thousands-

worldwide-take-part-in-largest-global-civil-disobedience-in-the-history-of-the-climate-movement/.

2.	United Steelworkers, "USW response to 350org's 'Break Free from Fossil Fuel' actions targeting oil refineries," April 26, 2016, http://www.usw.org/news/media-center/articles/2016/usw-response-to-350-orgs-break-free-from-fossil-fuel-actions-targeting-oil-refineries.

3.	Brian Nearing, "Rancor, protests greet top energy official," *timesunion*, May 11, 2016, http://www.timesunion.com/business/article/Protestors-disrupt-power-plant-owners-Colonie-7462672.php.

4.	"Thousands Converged in Albany to Blockade Bomb Trains," press release, http://www.albany2016.org/wp-content/uploads/2016/05/May14PressRelease.pdf; and personal observation.

5.	Lindsay Ellis, "Albany protest: 5 arrested after oil train delayed," *Albany Times-Union*, May 16, 2016, http://www.timesunion.com/local/article/Albany-Activists-gather-for-Break-Free-From-7468688.php.

6.	E-mail message, May 5, 2016.

7.	See Jeremy Brecher, *Climate Insurgency: A Strategy for Survival* (Boulder: Paradigm Publishers, 2015), updated 2016 edition, available for free download at www.jeremybrecher.org.

Chapter 2: Paris: The Promise of Betrayal

1.	"Projected growth in CO_2 emissions driven by countries outside the OECD," *Today in Energy*, May 6, 2016, http://www.eia.gov/todayinenergy/detail.cfm?id=26252&src=email.

2.	Coral Davenport, "Nations Approve Landmark Climate Accord in Paris," *New York Times*, December 12, 2015, http://www.nytimes.com/2015/12/13/world/europe/climate-change-accord-paris.html?_r=0.

3.	"Scoreboard Science and Data," *Climate Interactive*, https://www.climateinteractive.org/tools/scoreboard/scoreboard-science-and-data/.

4.	Andrew Jones et al., "With Improved Pledges Every Five Years, Paris Agreement Could Limit Warming Below 2C," *Climate Interactive*, December 14, 2015, https://www.climateinteractive.org/blog/press-release-with-an-ambitious-review-cycle-offers-to-paris-climate-talks-could-limit-warming-below-2c/.

5.   James Hansen, "Young People's Burden," *Climate Science, Awareness and Solutions*, October 4, 2016, http://csas. ei.columbia.edu/2016/10/04/young-peoples-burden/.

6.   Mark Hertsgaard, "Breakthrough in Paris," *The Nation*, January 4, 2016, quoting "a delegate from a Mediterranean country" who requested anonymity "because his government is a US ally."

7.   Andrew Restuccia, "The one word that almost sank the climate talks," *Politico*, December 12, 2015, http://www.politico. com/story/2015/12/paris-climate-talks-tic-toc-216721.

8.   Jones et al., "With Improved Pledges Every Five Years," https://www.climateinteractive.org/blog/press-release-with-an-ambitious-review-cycle-offers-to-paris-climate-talks-could-limit-warming-below-2c/.

9.   Davenport, "Nations Approve Landmark Climate Accord in Paris."

10.   Kim Nicholas, "Top scientists weigh in on current draft of Paris climate agreement," *Road to Paris*, December 11, 2015, http://roadtoparis.info/2015/12/11/top-scientists-weigh-in-on-current-draft-of-paris-climate-agreement/.

11.   United Nations Framework Convention on Climate Change, "Adoption of the Paris Agreement," December 12, 2015, http://unfccc.int/resource/docs/2015/cop21/eng/l09.pdf.

Chapter 3: The Power of the People vs. the Forces of Doom

1.   Gene Sharp, *The Politics of Nonviolent Action: Part One: Power and Struggle* (Boston: Porter Sargent, 1973), 32.

Chapter 4: A Strategic Vision

1.   For a fuller assessment of these bases of power, see Brecher, *Climate Insurgency*, www.jeremybrecher.org.

2.   Bill Moyer, *Doing Democracy* (Gabriola Island, BC: New Society Publishers, 2001), 16.

Chapter 5: Imposing a Fossil Freeze

1.   Oil Change International, "The Sky's Limit: Why the Paris Climate Goals Require a Managed Decline of Fossil Fuel Production," September 2016, http://priceofoil.org/content/uploads/2016/09/OCI_the_skys_limit_2016_FINAL_2.pdf; for an earlier fossil fuel freeze proposal, see Jeremy Brecher, "Freezing the Greenhouse," *Z*, January 8, 2010, https://

zcomm.org/zcommentary/freezing-the-greenhouse-the-snowball-strategy-by-jeremy-brecher/.

2. Bill McKibben, "Climate fight won't wait for Paris: vive la résistance," *Guardian*, March 9, 2015, http://www.theguardian.com/environment/2015/mar/09/climate-fight-wont-wait-for-paris-vive-la-resistance.

3. Katie Herzog, "Portland, being Portland, says no to new fossil fuel infrastructure," *Grist*, November 13, 2015, http://grist.org/article/portland-being-portland-says-no-to-new-fossil-fuel-infrastructure/.

4. Elizabeth Douglass, "In Keystone Fight and Beyond, Infrastructure Is Energy Policy," *Inside Climate News*, January 8, 2015, https://insideclimatenews.org/news/20150108/keystone-fight-and-beyond-infrastructure-energy-policy.

5. Gene Sharp, *The Politics of Nonviolent Action: Part Three, The Dynamics of Nonviolent Action* (Boston: Porter Sargent 1973), 471–72.

6. Jeremy Brecher, *Jobs Beyond Coal*, Labor Network for Sustainability, http://report.labor4sustainability.org/coal_2012.pdf.

Chapter 6: Imposing Climate Action Plans

1. First Amended Complaint for Declaratory and Injunctive Relief; Case No.: 6:15-cv-01517-TC, Filed 9/10/15, United States District, District of Oregon – Eugene Division, https://static1.squarespace.com/static/571d109b04426270152febe0/t/57a35ac5ebbd1ac03847eece/1470323398409/YouthAmendedComplaintAgainstUS.pdf.

2. First Amended Complaint, 85. If emissions had peaked and reductions had begun in 2005, only a 3.5% per year global reduction would have been necessary to reach 350 ppm by 2100.

3. See John Humphries, "Connecticut's Climate Change Mitigation Goals: Charting a Course to 2050," *Connecticut Roundtable on Climate and Jobs*, July 15, 2015, http://www.ct.gov/deep/lib/deep/climatechange/gc3/member_communications/humphries_2015_0717b.pdf.

4. Jones et al., "With Improved Pledges Every Five Years," https://www.climateinteractive.org/blog/press-release-with-an-ambitious-review-cycle-offers-to-paris-climate-talks-could-limit-warming-below-2c/.

5. Ibid.

Chapter 7: Self-Organization for Climate Defense

1.   For further discussion, see Jeremy Brecher, *Save the Humans?*, *Common Preservation in Action* (Boulder, CO: Paradigm Publishers, 2012), Chapter 47, "Power and Dependence."

2.   Anthony Leiserowitz et al., *Americans' Action to Limit Global Warming*, Yale Program on Climate Change Communication, http://climatecommunication.yale.edu/wp-content/uploads/2014/02/Behavior-November-2013.pdf, 5.

3.   Ibid., 14.

Chapter 8: Turning the Public against Fossil Fuels

1.   This analysis draws on Connie Roser-Renouf et al., "The genesis of climate change activism: from key beliefs to political action," *Climate Change*, July 2014, 5, 17.

2.   There is a vast literature regarding public opinion about climate change. I have used three studies based on reputable public opinion polling techniques: Anthony Leiserowitz et al., *Politics and Global Warming*, Yale Program on Climate Change Communication, Fall 2015, http://climatecommunication.yale.edu/wp-content/uploads/2016/01/Politics-and-Global-Warming-Fall-2015-2.pdf, surveys the opinions of registered voters on climate change; Anthony Leiserowitz et al., *Americans' Action to Limit Global Warming*, Yale Program on Climate Change Communication, http://climatecommunication.yale.edu/wp-content/uploads/2014/02/Behavior-November-2013.pdf, focuses on climate change action; Connie Roser-Renouf et. al, "The genesis of climate change activism: from key beliefs to political action," *Climate Change*, 125, no. 2 (July 2014) 163–78, examines the factors that affect climate change activism.

    Such poll data requires care in use. Poll data is generally not completely accurate. It focuses on the opinion of individuals in isolation rather than in interaction with others. The phrasing of questions and the categorization of the answers can distort the results. It can be difficult to distinguish short-term fluctuations from deep changes that occur over time. In this analysis I examine broad patterns and trends, not small differences or short-term fluctuations. My analysis focuses on possibilities for transformation of action, rather than static current opinion. One result is that it avoids fixed categories like the "six Americas" analysis.

3. Leiserowitz et al., *Politics and Global Warming*, 21.
4. Leiserowitz et al., *Americans' Actions to Limit Global Warming*, 17.
5. Leiserowitz et al., *Politics and Global Warming*, 22.
6. Ibid., 24.
7. Ibid., 18.
8. Ibid., 5–6.
9. Ibid., 20.
10. Ibid., 6.
11. Ibid.
12. Ibid., 15.
13. Ibid., 5.
14. Ibid., 5–6.
15. Giovanni Russonello, "Two-Thirds of Americans Want U.S. to Join Climate Change Pact," *New York Times*, November 30, 2015, https://www.nytimes.com/2015/12/01/world/americas/us-climate-change-republicans-democrats.html.
16. Leiserowitz et al., *Politics and Global Warming*, 15.
17. Ibid., 5, 23.
18. Roser-Renouf et al., "The genesis of climate change activism," 18.
19. Ibid., 18.

Chapter 9: Turning Climate Worriers into Climate Warriors

1. Anthony Leiserowitz et al., *Americans' Action to Limit Global Warming*, Yale Program on Climate Change Communication, http://climatecommunication.yale.edu/wp-content/uploads/2014/02/Behavior-November-2013.pdf, 5.
2. Ibid., 4.
3. Ibid., 5.
4. Ibid., 4–5.
5. Ibid., 13.
6. Ibid., 5.
7. Ibid., 17.
8. John Immerwahr, "Waiting for a Signal: Public Attitudes toward Global Warming, the Environment and Geophysical Research," *Public Agenda*, April 15, 1999, http://research.policyarchive.org/5662.pdf.
9. Connie Roser-Renouf et. al, "The genesis of climate change activism: from key beliefs to political action," *Climate Change* 125, no. 2, (July 2014): 176.
10. Ibid., 4.

11. Ibid., 17.

Chapter 10: Undermining the Pillars of Support for Climate Destruction

1. For a journalistic investigation of these activities, see Ross Gelbspan, *The Heat Is On* (New York: Perseus, 1997).

2. See the Business Environmental Leadership Council page on the Center for Climate and Energy Solutions website, http://www.c2es.org/business/belc.

3. "The U.S Chamber Doesn't Speak for Me," http://chamber.350.org/.

4. "Stop Heartland's Climate Denial!," http://act.350.org/sign/heartland.

5. Jeremy Brecher, "Making the Promises Real: Labor and the Paris Climate Agreement," *Common Dreams*, May 22, 2016, http://www.commondreams.org/views/2016/01/20/making-promises-real-labor-and-paris-climate-agreement.

6. Tik Root, "An Evangelical Movement Takes on Climate Change," *Newsweek*, March 9, 2016, http://www.newsweek.com/2016/03/18/creation-care-evangelical-christianity-climate-change-434865.html.

7. Anthony Leiserowitz et al., *Politics & Global Warming*, Yale Program on Climate Change Communication, Fall 2015, http://climatecommunication.yale.edu/wp-content/uploads/2016/01/Politics-and-Global-Warming-Fall-2015-2.pdf, 24.

8. Ibid., 18.

9. Ibid., 5–6.

10. Ibid., 9.

11. Emma Foehringer Merchant, "The 2016 presidential debates all but ignored climate change," *Grist*, October 19, 2016, http://grist.org/election-2016/climate-airtime-presidential-debate/.

12. For an example of such an accountability strategy, see Rachel Mandelbaum, "Report Card Warns Gov. McCauliffe is close to 'flunking out' on Climate and Clean Energy Priorities," *Chesapeake Climate Action Network*, http://chesapeakeclimate.org/press-releases/report-card-warns-gov-mcauliffe-is-close-to-flunking-out-on-climate-and-clean-energy-priorities/; for the electoral impact of climate activists, see, for example, Trip Gabriel and Coral Davenport, "'Fractivists' Increase Pressure on Hillary Clinton and Bernie

Sanders in New York," *New York Times*, April 4, 2016, http://www.nytimes.com/2016/04/05/us/politics/hillary-clinton-bernie-sanders-climate-change.html?_r=0.

13.   Michael Wolkind, "How we won acquittal of Kingsnorth six," *Guardian*, May 31, 2009, http://www.theguardian.com/environment/cif-green/2009/may/31/kingsnorth-defence-lawyer.

14.   Alec Johnson, "Blockadia on Trial: What the Jury Did Not Hear," *Huffpost Green*, January 11, 2015, http://www.huffingtonpost.com/alec-johnson/keystone-xl-blockade_b_6134732.html.

15.   Jeremy Brecher, "Climate activist argues resistance is necessary to protect the public trust," *Waging Nonviolence*, October 17, 2014, http://wagingnonviolence.org/2014/10/public-trust-goes-trial/.

16.   Sarah Lazare, "Not Guilty: Flood Wall Street Protesters Vindicated by Manhattan Court," *Common Dreams*, March 6, 2015, http://www.commondreams.org/news/2015/03/06/not-guilty-flood-wall-street-protesters-vindicated-manhattan-court.

Chapter 11: Just Transitions

1.   See Jeremy Brecher, Ron Blackwell, and Joe Uehlein, "If Not Now, When? A Labor Movement Plan to Address Climate Change," *New Labor Forum*, Fall 2014, http://www.labor4sustainability.org/wp-content/uploads/2014/09/NLF541793_REV1.pdf.

2.   "The Clean Energy Future: Protecting the climate, creating jobs, and saving money," Labor Network for Sustainability, 350.org, and Synapse Energy Economics, http://www.labor4sustainability.org/wp-content/uploads/2015/10/cleanenergy_10212015_main.pdf, based on research by a team led by Frank Ackerman of Synapse Energy Economics; for examples of what this would mean in particular states, see the Labor Network for Sustainability "Climate, Jobs, and Justice" project, http://climatejobs.labor4sustainability.org.

3.   "Fair Development Victory!," *We Demand Fair Development*, March 26, 2015, https://stoptheincinerator.wordpress.com; for examples of cooperation among environmental, labor, and community groups in the context of coal plant shutdowns, see "Jobs Beyond Coal: A Manual for Communities, Workers, and Environmentalists," http://report.labor4sustainability.

org; for "just transition," see Jeremy Brecher, "A Superfund for Workers: How to Promote a Just Transition and Break Out of the Jobs vs. Environment Trap," *Dollars & Sense*, November/ December 2015, http://www.labor4sustainability.org/wp-content/uploads/2015/10/1115brecher.pdf; see also the "Clean Energy Worker Just Transition Act" recently outlined by Sen. Bernie Sanders (I-Vt.), Jeff Merkley (D-Ore.), and Edward Markey (D-Mass.), http://www.sanders.senate.gov/ download/worker-just-transition-act-summary?inline=file.

4.   Kristen Sheeran et al., "The Keystone Pipeline Debate: An Alternative Job Creation Strategy," Economics for Equity and Environment and Labor Network for Sustainability, http:// www.labor4sustainability.org/files/__kxl_main3_11052013.pdf.

5.   Break Free Midwest, "Why We March!," https://midwest. breakfree2016.org/whywemarch/.

6.   Sen. Bernie Sanders (I-Vt.), Jeff Merkley (D-Ore.), Edward Markey (D-Mass.), "The Clean Energy Worker Just Transition Act," http://www.sanders.senate.gov/download/ worker-just-transition-act-summary?inline=file.

7.   For one example of such a broader movement and program, see Jeremy Brecher, *Save the Humans? Common Preservation in Action* (Boulder CO: Paradigm Publishers, 2012), Part 5, "Human Preservation."

Chapter 12: The Right of the People to Protect the Climate

1.   This chapter draws on Jeremy Brecher and David Solnit, "Using the 'Public Trust' to Frame 'Break Free From Fossil Fuels' Actions," *Common Dreams*, April 28, 2016, http://www.commondreams.org/views/2016/04/28/ using-public-trust-frame-break-free-fossil-fuels-actions.

2.   Our Children's Trust, "Securing the Legal Right to a Stable Climate," https://www.ourchildrenstrust.org.

Chapter 13: Dual Power

1.   Jeremy Brecher, "Climate destruction in the court of public opinion," *Waging Nonviolence*, October 2, 2014, http:// wagingnonviolence.org/feature/climate-destruction- court-public-opinion/; the live stream of the Climate Justice Tribunal is available at http://new.livestream.com/ TheNewSchool/peoples-climate-justice-summit.

2.   "Youths Secure Second Win In Washington State Climate Lawsuit," Our Children's Trust press release, April 29, 2016,

http://westernlaw.org/article/youths-secure-second-win-washington-state-climate-lawsuit-press-release-42916.

3.   "First Amended Complaint for Declaratory and Injunctive Relief," Case No.: 6:15-cv-01517-TC, Filed 9/10/15, United States District, District of Oregon – Eugene Division, https://static1.squarespace.com/static/571d109b04426270152febe0/t/57a35ac5ebbd1ac03847e e ce/1470323398409/YouthAmendedComplaintAgainstUS.pdf, 4–5.

4.   The examples in this section are discussed more fully in Jeremy Brecher, "A new wave of climate insurgents defines itself as law-enforcers," *Waging Nonviolence*, February 29, 2016, http://wagingnonviolence.org/feature/break-free-from-fossil-fuels-public-trust-domain/.

5.   "Press Release: Pennsylvania Township Legalizes Civil Disobedience," May 3, 2016, http://celdf.org/2016/05/press-release-pennsylvania-township-legalizes-civil-disobedience/.

6.   Jeremy Brecher and Tim Costello, *Common Sense for Hard Times* (Washington, DC: Two Continents/IPS, 1976), 206; see also Gene Sharp's discussion of "dual sovereignty and parallel government" as a form of nonviolent action in *The Politics of Nonviolent Action: Part Two, The Methods of Nonviolent Action*, 422–33.

Conclusion: Two Scenarios

1.   "Future Climate Change," "Projected Atmospheric Greenhouse Gas Concentrations" graph, EPA website, https://www3.epa.gov/climatechange/science/future.html.

2.   Ibid.

3.   Eric Holthaus, "New U.N. Report: Climate Change Risks Destabilizing Human Society," *Slate*, March 30, 2014, http://www.slate.com/blogs/future_tense/2014/03/30/ipcc_2014_u_n_climate_change_report_warns_of_dire_consequences.html.

4.   "Human Tide: The Real Migration Crisis," *Christian Aid Report*, May 2007, https://www.christianaid.org.uk/Images/human-tide.pdf.

5.   HM Treasury, "Stern Review on the Economics of Climate Change," http://webarchive.nationalarchives.gov.uk/+/http://www.hm-treasury.gov.uk/sternreview_index.htm.

About the Author

JEREMY BRECHER IS THE AUTHOR OF MORE THAN A DOZEN books on labor and social movements, including *Save the Humans? Common Preservation in Action* and his classic labor history *Strike!*, recently published in a revised and expanded fortieth anniversary edition by PM Press. He has been writing about climate protection since 1988, most recently in his book *Climate Insurgency: A Strategy for Survival* (2015). He holds a PhD from the Union Graduate School and is a cofounder of the Labor Network for Sustainability. In addition to being honored with five regional Emmy Awards for his documentary film work, Jeremy was arrested in the early White House sit-ins against the Keystone XL pipeline.

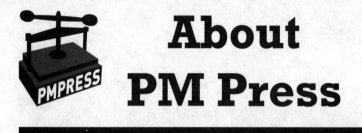

About
PM Press

politics • culture • art • fiction • music • film

PM Press was founded at the end of 2007 by a small collection of folks with decades of publishing, media, and organizing experience. PM Press co-conspirators have published and distributed hundreds of books, pamphlets, CDs, and DVDs. Members of PM have founded enduring book fairs, spearheaded victorious tenant organizing campaigns, and worked closely with bookstores, academic conferences, and even rock bands to deliver political and challenging ideas to all walks of life. We're old enough to know what we're doing and young enough to know what's at stake.

We seek to create radical and stimulating fiction and nonfiction books, pamphlets, T-shirts, visual and audio materials to entertain, educate, and inspire you. We aim to distribute these through every available channel with every available technology, whether that means you are seeing anarchist classics at our bookfair stalls; reading our latest vegan cookbook at the café; downloading geeky fiction e-books; or digging new music and timely videos from our website.

Contact us for direct ordering and questions about all PM Press releases, as well as manuscript submissions, review copy requests, foreign rights sales, author interviews, to book an author for an event, and to have PM Press attend your bookfair:

PM Press • PO Box 23912 • Oakland, CA 94623
510-658-3906 • info@pmpress.org

Buy books and stay on top of what we are doing at:

www.pmpress.org

FOPM
MONTHLY SUBSCRIPTION PROGRAM

These are indisputably momentous times—the financial system is melting down globally and the Empire is stumbling. Now more than ever there is a vital need for radical ideas.

In the many years since its founding—and on a mere shoestring—PM Press has risen to the formidable challenge of publishing and distributing knowledge and entertainment for the struggles ahead. With over 200 releases to date, we have published an impressive and stimulating array of literature, art, music, politics, and culture. Using every available medium, we've succeeded in connecting those hungry for ideas and information to those putting them into practice.

Friends of PM allows you to directly help impact, amplify, and revitalize the discourse and actions of radical writers, filmmakers, and artists. It provides us with a stable foundation from which we can build upon our early successes and provides a much-needed subsidy for the materials that can't necessarily pay their own way. You can help make that happen—and receive every new title automatically delivered to your door once a month—by joining as a Friend of PM Press. And, we'll throw in a free T-Shirt when you sign up.

Here are your options:
- $30 a month: Get all books and pamphlets plus 50% discount on all webstore purchases
- $40 a month: Get all PM Press releases (including CDs and DVDs) plus 50% discount on all webstore purchases
- $100 a month: Superstar—Everything plus PM merchandise, free downloads, and 50% discount on all webstore purchases

For those who can't afford $30 or more a month, we have **Sustainer Rates** at $15, $10 and $5. Sustainers get a free PM Press T-shirt and a 50% discount on all purchases from our website.

Your Visa or Mastercard will be billed once a month, until you tell us to stop. Or until our efforts succeed in bringing the revolution around. Or the financial meltdown of Capital makes plastic redundant. Whichever comes first.

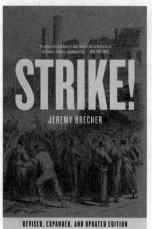

Strike!
Revised and Expanded
Jeremy Brecher
$24.95 • ISBN: 978-1-60486-428-1

Since its original publication in 1972, no book has done as much to bring American labor history to a wide audience. *Strike!* narrates the dramatic story of repeated, massive, and sometimes violent revolts by ordinary working people in America. It tells this exciting hidden history from the point of view of the rank-and-file workers who lived it.

This expanded edition brings the story up to date, covering the forty years since the original placed the problems faced by working people today in the context of 140 years of labor history.

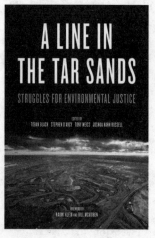

A Line in the Tar Sands
Struggles for Environmental Justice
Edited by Joshua Kahn Russell, Stephen D'Arcy, Tony Weis, and Toban Black
Foreword by Naomi Klein and Bill McKibben
$24.95 • ISBN: 978-1-62963-039-7

The fight over the tar sands in North America is among the epic environmental and social justice battles of our time, and one of the first that has managed to quite explicitly marry concern for frontline communities and immediate local hazards with fear for the future of the entire planet.

Including leading voices involved in the struggle against the tar sands, *A Line in the Tar Sands* offers a critical analysis of the impact of the tar sands and the challenges opponents face in their efforts to organize effective resistance.